Microplastics in Human Consumption

This book highlights plastic pollution, especially microplastics, and the consequences on the environment and on the human diet through food chain/web transmission and contamination through processing and packaging of various human food items including water and beverages. This book provides a detailed outlook on microplastics, their origin, distribution, categories, presence in human food items and impact on the ecosystem, organisms and so forth. It covers updated information on microplastic contamination in the biotic and abiotic products of the sea/ocean, the contamination of microplastics in drinking water/bottled water, honey, sugar and so forth.

Features:

- Discusses the presence of microplastics in matrices prone to human consumption.
- Includes general information on microplastics, their origin, types, shapes, size and nomenclature.
- Reviews microplastics in different types of human consumable items.
- Illustrates fundamental methods and techniques used in microplastics research.
- Explores overall impacts of microplastics in the organisms contaminated with the same.

This book is aimed at researchers, graduate students and faculty members interested in plastic pollution, microplastics, marine litter, environmental science, food consumption and analysis.

Microplastics in Human Consumption

E. V. Ramasamy and Ajay Kumar Harit

CRC Press
Taylor & Francis Group
Boca Raton London New York

CRC Press is an imprint of the
Taylor & Francis Group, an **informa** business

First edition published 2023
by CRC Press
6000 Broken Sound Parkway NW, Suite 300, Boca Raton, FL 33487-2742

and by CRC Press
4 Park Square, Milton Park, Abingdon, Oxon, OX14 4RN

CRC Press is an imprint of Taylor & Francis Group, LLC

© 2023 E. V. Ramasamy and Ajay Kumar Harit

ISBN: 978-1-032-06325-6 (hbk)
ISBN: 978-1-032-06327-0 (pbk)
ISBN: 978-1-003-20175-5 (ebk)

DOI: 10.1201/9781003201755

Typeset in Times New Roman
by Deanta Global Publishing Services, Chennai, India

Dedicated to

Beloved parents

Appa Thiru E. Venkatesaperumal and *Amma* Thirumathi V. Jayalakshmi

-E. V. Ramasamy

To my parents

(Mr. M.L Harit and Mrs. Ranjeet Kaur),

and

my siblings

for their encouragement and support of all my endeavours

-Ajay Kumar Harit

Contents

Foreword

Mahatma Gandhi University

Priyadarsini Hills P.O., Kottayam, Kerala, India, 686560
Website: www.mgu.ac.in
(Established by Kerala State Legislature by Notification No.3431/Leg. Cl/85/
Law, dated 17th April 1985)

Vice-Chancellor   June 6, 2022

Anthropogenic activities have been identified as the most significant drivers
of environmental degradation. The widespread occurrence of plastic debris is
a characteristic marker of human interference with the environment. Plastics,
due to their intrinsic qualities like malleability, durability, resistance to degradation and low cost, have become ideal materials for a variety of industrial
and consumer products. An exponential increase in the production and usage
of plastics has resulted in the accumulation of plastic waste, posing serious
environmental and disposal issues. Microplastic pollution is one such issue of
plastic litter being researched globally in recent years. Microplastics (plastic
pieces of < 5 mm) and nanoplastics (< 1 μm) are formed due to the fragmentation of large plastic debris or are synthesized for various industrial purposes
and discarded as such into the environment after use. Global research on this
topic has tremendously increased in recent years with much focus on marine
microplastics while freshwater and terrestrial environments are studied less.
One of the serious issues with microplastics in aquatic environments is that
in addition to their inbuilt chemicals such as additives, colorants, phthalates,
etc., they do adsorb many pollutants including pesticides and heavy metals
from the surrounding aquatic environment. When microplastics with all these
toxic substances are consumed by organisms followed by the trophic transfer
through the food chain and food web, it becomes a potential threat to human

health as dietary items such as seafood and sea salt are contaminated with microplastics. Infiltration of microplastic into human consumable items has attracted global concern. In this context, the authors have described the microplastic contamination reported in major human dietary items such as fishes, clams, sea salt, sugar, bottled water, etc. in this book. The method of sampling, extraction and analysis of microplastics from such food items has been lucidly presented in the book.

Dr. Ramasamy is known to me as a faculty member of this university who has initiated microplastics research at MG University and has good publications on microplastics. Dr. Harit (co-author) joined this university as a national post doctoral fellow (N-PDF) under the mentorship of Dr. Ramasamy and also has good contributions on microplastics. To my understanding this book becomes significant as it is written on a nascent but highly relevant topic: microplastics in human consumption; perhaps this book is the first of its kind being published in India. The book is suitable for students, researchers and scientists who are beginners in microplastics research. Essentially, this book is exactly what it says on the cover: *Microplastics in Human Consumption*. The topic is, after all, an extremely important one as it looks into various aspects of human negligence regarding the use of plastics. Since the matters pertaining to microplastics in human consumption are still relatively less attended ones, it is very tempting to presume that the topics this book covers are completely new – or at least significantly different – as well. I am happy that the effort taken by Dr. Ramasamy and Dr. Harit in writing this book has not gone in vain. As such, the book is an essential read for everyone with a broad interest in the ethics of the environment. I congratulate both the authors and wish the book a great reach and success.

Prof. Sabu Thomas

Preface

"Dawn of the Plasticene", an article by Christina Reed that appeared in *New Scientist* (2015), highlights that "Our Love for plastics is leaving a lasting legacy". The following are some of the key points of the article:

> 5.25 trillion pieces of plastics are currently floating at sea; 80 per cent of marine litter comes from land; plastic particles also act as sponges and absorb organic pollutants, pesticides in sea water and such particles can poison the fish; plastic particles in ocean are providing an entirely new ecosystem, the "plastisphere", where microbes including pathogens like *Vibrio* can colonise over these particles; marine microplastics could pose a threat to food safety.

This article has inspired me toward microplastics research. Along with my research team, I started my initial work on assessing the microplastic pollution in the sediment of Vembanad Lake, one of the three Ramsar sites in Kerala, India. Our findings from this research became the first report from India on the occurrence of microplastics in the sediment of a lake. Following this, our team studied the presence of microplastics in the water and sediment of rivers, estuaries, polar water bodies like Arctic fjords and the Southern Ocean (Antarctic water), as well as road dust, indoor dust, etc. Of late, we realized the significance of the microplastics when they enter the food chain/food web and the risk of trophic-level transfer leading to microplastic contamination of human food items; we then started working on microplastic contamination in human consumable items, which became the starting point of this book. Vast literature available on this subject formed the basis of this book besides our own findings on microplastics in seafood items like mussels, sea salt and potable water. We thank the authors of all publications on plastics and microplastics available at the time of writing this book.

When I was onboard of the SA *Agulhas* (South African research vessel) as a participant of the Indian Scientific Expedition to the Southern Ocean (ISESO 2020) on my polar research project on microplastics, my postdoctoral fellow Dr. Ajay Kumar Harit (co-author of this book) conveyed me the news of acceptance of our book proposal by Taylor & Francis (CRC Press) over a WhatsApp call as my access to email was very limited onboard; thus we started working on the chapters of this book while the *Agulhas* was sailing toward Antarctica over the roaring 40s and screaming 60s of the Southern Ocean.

In this book, we have provided an overview of plastics and microplastics in the first chapter. Different categories and classifications based on size, shape and polymer content of microplastics have been explained in this chapter. The following two chapters deal with the microplastic contamination of human food items, potable water and beverages. The methods used in microplastics extraction, analysis and polymer identification and general impacts of microplastics have been presented in the subsequent chapters. Management aspects, legislation, policy and mitigation strategies are provided in the last chapter. In summary this book, apart from serving as a first reading on microplastic contamination of human food items for students, is also ideal for beginners in microplastic research as it has a good compilation of methods being used in the extraction, analysis and polymer identification of microplastics.

My research team working on microplastics has also helped us in making this book; the team includes Dr. Sruthi S. Nair, Dr. Arun Babu, Dr. Vipin Joseph Marcose, Ms. Sruthy S., Mr. Naveen Babu, Ms. Anagha P. L., Ms. Sandya K. and Ms. Viji N. V. We are grateful to the National Center for Polar and Ocean Research (NCPOR), Goa, for their support on the Arctic and Southern Ocean expeditions; similarly, we are also thankful to the National Center for Coastal Research (NCCR), the Ministry of Earth Sciences (MoES), Government of India and Kerala State Council for Science, Technology and Environment (KSCSTE), Thiruvananthapuram, for the financial support in conducting microplastics research. Dr. Ajay Kumar Harit acknowledges the financial support of DST-SERB for the N-PDF. We thank Mahatma Gandhi University, Kottayam, for all the support, infrastructure and instrumentation facilities including the DST-SAIF facility for Raman spectroscopic analysis of the microplastics. I thank my wife Dr. P. Padma and my son R. Kesavapugazhendhi for their encouragement and support in writing this book.

It has been a pleasure to work with Taylor & Francis (CRC Press) and we are looking forward to associating with them in the future.

– Prof. E. V. Ramasamy

Authors

 **E. V. Ramasamy** is a Professor at the School of Environmental Sciences, Mahatma Gandhi University, Kottayam, Kerala, India. He is also the Dean of the Faculty of Environmental and Atmospheric Sciences and the former Director of the School of Environmental Sciences. In his 30 years of professional career, he has guided 16 PhD scholars and 30 MPhil scholars and has 120 research publications including three books to his credit. He is also serving as a member of several committees of the Government of Kerala such as the State Waste Management Policy Committee and the Technical Committee of Suchitwa Mission, Local Self Government Department (LSGD) as well as the Kerala State Council for Science, Technology and Environment (KSCSTE) – Programme Advisory Committee for Environment and Ecology Programme, Government of Kerala. He has also participated in the Indian Arctic Expedition (Summer batch 2019) and the Indian Scientific Expedition to the Southern Ocean (Antarctic waters) ISESO 2020; in both these expeditions he has done research on microplastics in the polar environment. He has been awarded the DUO-India 2020 Professor Fellowship Award by the ASEM-DUO Fellowship Program (Seoul, South Korea).

He works on municipal solid waste (MSW) management, wastewater treatment, microbial fuel cells (MFCs), water pollutants like heavy metals, landfill leachates and microplastics. His work on earthworms and soil carbon dynamics has been published in *Scientific Reports* of the Nature publishing group. His paper on microplastics in the sediments of Vembanad Lake was the first report from India, and the recent publication on microplastics in the Kongsfjord, Arctic, was the outcome of his polar research published in 2021.

 **Ajay Kumar Harit** has served as a Research Associate at the School of Environmental Sciences, Mahatma Gandhi University, Kottayam, Kerala, India. His doctorate is from Pondicherry University (a Central University), Puducherry, India. He did his postdoctoral work at the Indo-French Cell for Water Sciences, Civil Engineering Department, Indian Institute of Science (IISc), Bangalore, during 2016–2017. He was awarded the DST-SERB-National Post-Doctoral Fellowship (N-PDF) in 2017; he joined Mahatma Gandhi University, Kottayam, Kerala, as an N-PDF in 2017 and completed the associateship successfully. His doctoral and postdoctoral research work was on exploring the attributes of termites (one of the ecosystem engineers) and their influence on the soil and forest ecosystem focusing on soil and nutrient dynamics.

During his stay at Mahatma Gandhi University, he developed his interest in microplastics research with a focus on monitoring microplastics contamination in water bodies and human consumable items. He has published research articles on microplastic pollution in the surface sediment of Kongsfjorden, Svalbard Arctic, in the *Marine Pollution Bulletin* (Elsevier) as co-author. A book chapter co-authored by him has been published by Cuvillier Verlag, Gottingen, Germany. He also presented his research findings at national and international conferences in the field of microplastics. He has 18 articles (of these 16 in SCI-indexed journals) and two book chapters published to his credit as the outcome of his doctoral and postdoctoral research.

Plastics and Microplastics in the Environment

1

1.1 INTRODUCTION

Plastic is a wonderful invention of mankind occupying almost all spheres of human life. Plastics have become inevitable companions in our daily life. We begin the day with a plastic toothbrush and paste; certain brands of toothpaste do contain microplastics (tiny plastic particles) and thus our daily cleaning and rinsing of our teeth contribute their share of microplastics to the environment. The discarded plastic waste and its management is the issue behind the ban on plastics, especially single-use plastics (SUP), being imposed by several nations. The lack of proper mechanisms for municipal solid waste (MSW) management often leads to the accumulation of waste piles with plastic debris; such dumps also become a major contributor of plastics to water bodies. The plastic wastes on reaching the water bodies or on the way may get disintegrated into smaller pieces; such fragmented plastics of less than 5 mm size are referred to as 'microplastics' (MPs) – a recent addition to the class of contaminants of emerging concern (CEC).

1.2 PLASTICS

Plastics are organic synthetic polymers, which have been used widely by mankind mainly due to their intrinsic properties like durability, inexpensiveness, being lightweight, etc. The history of plastic evolution – rather revolution – is, however, recent. Bakelite was the first plastic synthesized in the 1920s; later plastics production increased during the Second World War in order to meet the demand from the military for manufacturing parachutes, ropes, components of aircraft, automobiles and helmets. After the World War, the demand for

DOI: 10.1201/9781003201755-1

plastics was from the industries manufacturing household goods. Since then the demand from the consumer goods sector has been steeply increasing as plastic articles became very popular due to their inherent qualities mentioned earlier. Thus, the annual production of plastics from 30 million tons in the 1970s has become more than 359 million tons in 2018 (Plastics Europe, 2019).

Besides the increase in plastic production, another change noticed in the plastic industry in the last few decades was a shift from the production of durable plastics to single-use plastics. Much of the single-use plastics (SUP) such as LDPE, HDPE, PS and EPS production is from Northeast Asia (China, Hong Kong, Japan, Republic of Korea and Taiwan) followed by North America, the Middle East and Europe (UNEP, 2018). Greater concern about this global change in the SUP production scenario is on one hand due to the consumption of fossil hydrocarbons, which are non-renewable resources, and the issues related to the mismanagement of plastic wastes, especially SUP discards. An increase in production and consumption of plastics proportionately increases the quantity of discards; a recent report indicates about 4.8–12.7 million tons of plastics wastes reach the ocean annually (Jambeck et al., 2015; Iñiguez et al., 2017), making oceans the largest waste bins of plastic discards.

1.2.1 Major Types of Plastics

Based on the thermal properties such as melting and cooling capacities, plastics are grouped into two categories: thermoplastics and thermosets.

1.2.1.1 Thermoplastics

Thermoplastics can be melted and hardened by heating and cooling respectively. In other words, such plastics can be reheated, reshaped and frozen repeatedly. The most common thermoplastics are high-density polyethylene (HDPE), polyethylene terephtalate (PET), polyvinyl-chloride (PVC), polypropylene (PE), low-density polyethylene (LDPE), expanded polystyrene (EPS), polystyrene (PS), polycarbonate (PC), polypropylene (PP), polylactic acid (PLA) and polyhydroxy alkenoates (PHA).

1.2.1.2 Thermosets

Thermosets are plastics that, on heating, undergo a chemical change and form a stable three-dimensional network; thus these plastics cannot be remelted and reformed, hence they are *"non-reversible plastic"*. Common examples of thermosets are polyurethane (PUR), epoxy resins, phenolic resins, urea-formaldehyde (UF) resins, silicone, vinyl ester and acrylic resins.

1.2.1.3 Plastic Pollution

Plastic pollution is a growing concern at the global level as it is ubiquitous, persistent and harmful to many ecosystems. Plastic wastes, unlike other human-generated organic garbage, persist in the environment for years, polluting the environment and affecting the organisms enormously (Ramasamy et al., 2019). About 80% of plastic debris in oceans originates from land and is transported to oceans through the network of water bodies including streams, rivers, lakes, wetlands and estuaries (Sruthy and Ramasamy, 2017).

Various studies have documented the occurrence of plastic debris in the marine environment (Barnes et al., 2009; Cózar et al., 2014; Eriksen et al., 2014; Van Sebille et al., 2015), in lakes (Eriksen et al., 2013; Free et al., 2014), on shorelines (Browne et al., 2011) and in rivers (McCormick et al., 2014; Klein et al., 2015; Lechner et al., 2014; Yonkos et al., 2014; Kooi et al., 2016). Recent estimates indicate that between 1.15 and 2.41 million tons of plastic waste are transported to seas annually by rivers (Lebreton et al., 2017) and this is expected to increase in the coming decades (Jambeck et al., 2015).

The production and consumption of plastics by human society and the mismanagement of plastic discards are the major causes of plastic pollution. It is estimated that close to five trillion plastic bags are being consumed worldwide each year (UNEP, 2018). Five trillion in a year is almost equal to ten million plastic bags consumed a minute. Due to their light weight, plastic bags are easily transported or blown in the air, eventually ending up on land or in the ocean. On reaching the sea or ocean, they may get fragmented – due to wave action, UV radiation or biological means – into smaller pieces of plastics, subsequently becoming microplastics when their size becomes less than 5 mm.

1.3 MICROPLASTICS

As per the National Oceanic and Atmospheric Administration (NOAA) of the USA, microplastics (MPs) are any kind of plastic fragment measuring less than five millimeters in length (Arthur et al., 2009; Collignon et al., 2014; Crawford and Quinn, 2016). Thompson et al. (2004) were the first to use the term 'microscopic plastics' for plastic debris which were of size around 20 μm; later, this range was widened and the term microplastics (MPs) is being used for plastic particles smaller than 5 mm in size (Arthur et al., 2009; Betts, 2008; Fendall and Sewell, 2009; Hidalgo-Ruz et al., 2012). Further, the plastic particles depending

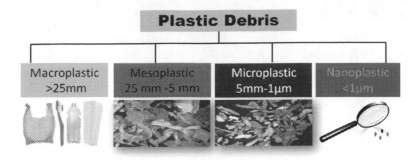

FIGURE 1.1   Classification of plastic debris on the basis of size.

on size have different nomenclature (Figure 1.1): macroplastics ($\geq$ 25mm); mesoplastics (5 to 25 mm); microplastics (1 μm to 5 mm); and nanoplastics (<1 μm) (Crawford and Quinn, 2016). Apart from size, MPs are also recognized based on different shapes they exhibit such as fragments, films, pellets, lines, fibers, filaments and granules (Driedger et al., 2015).

1.3.1 Categories of Microplastics

Microplastics enter into the environment either as primary or secondary microplastics.

1.3.1.1 Primary Microplastics

Primary microplastics are synthesized plastic particles of microscopic size often referred to as pellets or beads for various applications. Resin pellets are used as raw materials in the plastic industry for the production of various plastic articles (Peng et al., 2017). Microbeads find their applications in cosmetics mainly as exfoliating materials, abrasive agents in detergents and air blast media; hygiene and personal care products such as soaps, hand and facial cleansers, toothpaste, shower gels and shampoos (Arthur et al., 2009; Cole et al., 2011; Fendall and Sewell, 2009; Siegfried et al., 2017; Peng et al., 2017). Despite their huge industrial and domestic applications, primary microplastics represent only a fraction of the global MPs pollution load. Moreover, their use in cosmetics and personal care products has been reduced in many countries by imposing strict regulations (Lassen et al., 2015). The major concern of MPs pollution is due to the secondary microplastics.

1.3.1.2 *Secondary Microplastics*

Secondary microplastics originate from fragmentation and disintegration of larger plastic materials mostly plastic wastes discarded. The breakdown of plastic debris may occur due to the sunlight (UV radiation) or the mechanical action of wind/waves or biological means (Wagner et al., 2018; Gewert et al., 2015; Siegfried et al., 2017). However, the rate at which the disintegration occurs is highly dependent on environmental conditions such as temperature, the amount of UV light available, whether the plastic is on the land surface exposed to sunlight or buried or in an ocean, whether it is floating or submerged, etc. (UNEP, 2017). The polymer type of plastic determines whether the plastic will float or sink in an aquatic environment. When the specific gravity of the polymer exceeds 1.02 then it will sink while the plastics with a specific gravity less than 1.02 (seawater specific gravity) will float in the marine environment (Table 1.1).

Beaches are an ideal environment for fragmentation of plastic debris due to wind action, exposure to UV and physical abrasion from wave action. Once plastics are buried in beach sand or submerged, UV light penetration and temperature

TABLE 1.1   Common Plastics (Polymers) Found in the Marine Environment, Their Use and Specific Gravity

Categories or classes	Common applications	Specific gravity	
Polyethylene (PE)	Plastic bags, six-pack rings, gear	0.91-0.94	
Polypropylene (PP)	Rope, bottle caps, gear, strapping	0.90-0.92	Float
Polystyrene (expanded) (PS)	Bait boxes, floats, cups	0.01-1.05	
Sea Water		~ 1.02	
Polystyrene (PS)	Utensils, containers	1.04-1.09	
Polyvinyl chloride (PVC)	Film, pipe, containers	1.16-1.30	
Polyamide or nylon	Gear, rope	1.13-1.15	Sink
Polyethylene terephthalate (PET)	Bottles, strapping, gear	1.34-1.39	
Polyester resin + glass fibres	Textiles	>1.35	

Sources: UNEP (2017); Andrady (2011).

are greatly reduced and fragmentation proceeds much more slowly. Similarly, in the ocean, the floating plastics also undergo faster disintegration than those submerged as both UV penetration and the temperature are less at the bottom of the ocean. In addition to the polymer type, the additives added to the plastics also have a role in the disintegration process. For instance, if the plastic has an additive such as UV stabilizers then the UV action on the disintegration of plastics may be slow (UNEP, 2017; Peng et al., 2017).

1.3.1.3 Microplastics Classification Based on Size and Shape

Based on the size, plastic fragments are classified as macroplastics when their size is ≥ 25 mm; mesoplastics if their size ranges between 5 mm and 25 mm; microplastics if the size is between 1 and 5 mm; mini-microplastics if the size is between < 1 mm and 1 μm and nanoplastics if the size is < 1 μm (Crawford and Quinn, 2016).

Apart from size, MPs are also recognized based on different shapes they exhibit (Figure 1.2) such as: spheres (beads, pellets and granules), most of which belong to the primary MP category as these are manufactured as such; fibers (filaments and lines), films, fragments and foams (Free et al., 2014; Karami et al., 2017), which enter the environment due to the fragmentation of larger plastic debris (secondary

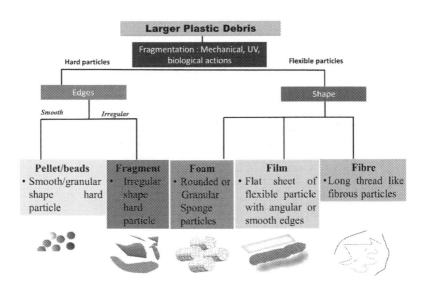

FIGURE 1.2 Schematic illustration of morphological form of microplastics. Source: Adopted from GESAMP (2019).

TABLE 1.2   Classification of Microplastics Based on Their Shape and Size

TYPE/SHAPE	SIZE	DESCRIPTION
Fragment (FR)	< 5 mm–1 mm	Plastic piece of irregular shape with a size of 1 mm to 5 mm along its longest dimension
Microfragment (MFR)	< 1 mm–1 μm	Plastic piece of irregular shape with a size of 1 μm to 1 mm along its longest dimension
Pellet (PT)	< 5 mm–1 mm	Small spherical-shaped plastic piece with a size of 1 mm to 5 mm in diameter
Microbead (MBD)	< 1 mm–1 μm	Small spherical-shaped plastic piece with a size of 1 μm to 1 mm in diameter
Fiber (FB)	< 5 mm–1 mm	Plastic piece of a strand or filament with a size of 1 mm to 5 mm along its longest dimension
Microfiber (MFB)	< 1 mm–1 μm	Plastic piece of a strand or filament with a size of 1 μm to 1 mm along its longest dimension
Film (FI)	< 5 mm–1 mm	Plastic piece of a thin membrane or sheet with a size of 1 mm to 5 mm along its longest dimension
Microfilm (MFI)	< 1 mm–1 μm	Plastic piece of a thin membrane or sheet with a size of 1 μm to 1 mm along its longest dimension
Foam (FM)	< 5 mm–1 mm	Plastic piece of foam or foam-like plastic material or sponge with a size of 1 mm to 5 mm along its longest dimension
Microfoam (MFM)	< 1 mm–1 μm	Plastic piece of foam or foam-like plastic material or sponge with a size of 1 μm to 1 mm along its longest dimension

MPs). Crawford and Quinn (2016) have made a detailed classification combining the size and shape of the plastic particles (Table 1.2).

1.4 SOURCE OF MICROPLASTICS

As far as the secondary microplastics are concerned, they reach the environment by means of being discarded after use or through accidental spill. The major source of secondary microplastics is the plastic discards or wastes disposed of unscientifically in the environment. Around 80% of microplastics found in the marine environment have terrestrial origin, transported by rivers, streams, estuaries and other means like run-off, sewage and industrial discharges and wastewater treatment plant effluents.

1.4.1 Microplastics in Terrestrial Environment

The main source of plastics (subsequently MPs) is anthropogenic activities. In this context, about 50% of the global population lives within 100 km of a coastline (UNEP, 2017). Accordingly, the quantity of plastic wastes entering the ocean from such land-based sources is expected to be high and likely to increase in future, unless significant changes are brought to the waste management practices on land. A recent estimate (Geyer et al., 2017) states that about 79% of the plastic wastes so far generated remains in landfills or solid waste dumps or in the environment, while 12% has been incinerated and only 9% has been recycled. This estimate gives a fair idea about the huge quantity of plastic waste remaining in the terrestrial environment with the potential of getting fragmented into MPs.

Microplastic pollution can originate from point or diffuse sources. The major point sources include domestic sewage and wastewater treatment plants (WWTPs) effluent and laundry effluents. MPs in the sewage are either being transported to a centralized wastewater treatment plants (WWTPs) or, as in many less developed countries (LDC), they are discharged untreated into adjacent water bodies. Similar is the case with laundry wastewater from individual houses; either manual wash or washing machine effluents carry MPs such as microfibers of polyester and nylon from cloths of synthetic fabric. Around 1,900 fibers per garment were reported from the effluent of a domestic washing machine (Browne et al., 2011). In the case of WWTPs effluents, there are contradicting reports on the MPs exiting through them. According to Conley et al. (2019) about 98% and 95% of MPs are removed from the wastewater by secondary and tertiary WWTPs respectively. On the contrary, substantial quantities of MPs have been reported from the WWTPs effluents; about 1,700 MPs per hour (Magnusson and Norén, 2014) and 65 million particles per day (Murphy et al., 2016, 2017) have been reported. As huge volumes of effluents are discharged globally by WWTPs, quite large quantities of MPs enter the environment from these sources despite high treatment efficiencies (Peng et al., 2017; He et al., 2019).

Non-point or diffuse sources are sources without any specific point of discharge of MPs. Surface run-off, rainfall or wind serve as major diffuse sources of MPs entry into any water body. Municipal solid waste dump yards – most of them ill-managed in many of the less developed countries – serve as sources of MPs when surface run-off or rainwater conduits the plastics into a nearby stream, river or lake.

1.4.2 Microplastics in Air Environment

Wind also acts as a dispersing means of plastic debris, mainly from the waste piles, dump yards and landfills (Barnes et al., 2009). The presence of MPs as fibers in atmospheric fallout, indoor and outdoor air has been reported (Dris et al., 2016, 2017). However, a number of studies reported the MPs in the air environment as less.

1.4.3 Microplastics in Aquatic Environment

It has been widely accepted that among all means of transport of MPs, canals, streams and rivers play the main role by connecting land to sea and act as an important pathway of microplastic waste generated on land to reach the marine environment. The key issues related to the presence of MPs in aquatic environments are due to the contamination of the food chain/web of the aquatic ecosystems. Due to their smaller size, MPs are often consumed by the organisms as feed; thus MPs enter the food chain/web and get transferred to higher trophic levels. Apart from the biotic components of the ecosystem, the abiotic products like salt from sea also get contaminated with MPs.

1.5 GENERAL IMPACTS OF MICROPLASTICS IN WATER BODIES

Given the longevity of MPs as hundreds and thousands of years, their presence in water bodies has been reviewed extensively in the literature. As far as the fate of MPs in sea/ocean waters is concerned, MPs are subjected to the following four mechanisms: sedimentation, shore deposition, nano-fragmentation and ingestion by organisms (Law and Thompson, 2014; Law et al., 2014; Cózar et al., 2014; Eriksen et al., 2014). When deposited on shore, a sufficient quantity of MPs on beaches could alter the physical properties of beach sediments (Carson et al., 2011), affecting crustaceans and sea turtles and their eggs. Floating MPs on the sea surface may interfere with light penetration which in turn will affect the organisms living beneath. The most important concern of risks of MPs in oceans and seas is due to the ingestion of MPs by a variety of organisms in the aquatic environment (Peng et al., 2017). Around 180 species including fish, turtles, birds and mammals have been reported to be ingesting MPs. The ingested MPs have been reported from different tissues and organs including gills, digestive glands, stomach and hepatopancreas (Von Moos et al., 2012; Farrell and Nelson, 2013). Translocation of MPs from the gut to the circulatory system to various cells and tissues of the blue mussel (*Mytilus edulis*) has been documented (Browne et al., 2008; Von Moos et al., 2012). Transfer of MPs from lower- to higher-trophic-level organisms/animals has been reported (Farrel and Nelson, 2013; Setälä et al., 2014). The greatest concern is that through this trophic-level transfer many of the sea food items used for human consumption may be contaminated with MPs.

As learned from the preceding sections of this chapter, microplastics – either as primary or secondary type – ultimately reach the marine environment.

The major impacts of the MPs in the aquatic environment have been briefly highlighted in this chapter. Apart from the physical and ecological impacts of MPs on the aquatic environment, their impact on the food chain/food web is more critical. Further chapters of this book focus on contamination of human food items with MPs either through food chain/web pathways or through flaws in the food industry while manufacturing, processing, packaging, etc.

REFERENCES

Andrady, A. L. (2011). Microplastics in the marine environment. *Marine pollution bulletin*, *62*(8),1596–1605.

Arthur, C., Baker, J. E., & Bamford, H. A. (Eds.) (2009). Proceedings of the International Research Workshop on the Occurrence, Effects, and Fate of Microplastic Marine Debris, September 9-11, 2008, University of Washington Tacoma, Tacoma, WA, USA, NOAA Technical Memorandum NOS-OR & R-30.

Barnes, D. K., Galgani, F., Thompson, R. C., & Barlaz, M. (2009). Accumulation and fragmentation of plastic debris in global environments. *Philosophical Transactions of the Royal Society B: Biological Sciences*, *364*(1526), 1985–1998.

Betts, K. (2008). Why small plastic particles may pose a big problem in the oceans. *Environmental Science and Technology*, *42*(24), 8995–8995.

Browne, M. A., Crump, P., Niven, S. J., Teuten, E., Tonkin, A., Galloway, T., & Thompson, R. (2011). Accumulation of microplastic on shorelines woldwide: Sources and sinks. *Environmental Science & Technology*, *45*(21), 9175–9179.

Browne, M. A., Dissanayake, A., Galloway, T. S., Lowe, D. M., & Thompson, R. C. (2008). Ingested microscopic plastic translocates to the circulatory system of the mussel, Mytilus edulis (L.). *Environmental Science & Technology*, *42*(13), 5026–5031.

Carson, H. S., Colbert, S. L., Kaylor, M. J., & McDermid, K. J. (2011). Small plastic debris changes water movement and heat transfer through beach sediments. *Marine Pollution Bulletin*, *62*, 1708–1713.

Cole, M., Lindeque, P., Halsband, C., & Galloway, T. S. (2011). Microplastics as contaminants in the marine environment: A review. *Marine Pollution Bulletin*, *62*(12), 2588–2597.

Collignon, A., Hecq, J. H., Galgani, F., Collard, F., & Goffart, A. (2014). Annual variation in neustonic micro-and meso-plastic particles and zooplankton in the Bay of Calvi (Mediterranean–Corsica). *Marine Pollution Bulletin*, *79*(1–2), 293–298.

Conley, K., Clum, A., Deepe, J., Lane, H., & Beckingham, B. (2019). Wastewater treatment plants as a source of microplastics to an urban estuary: Removal efficiencies and loading per capita over one year. *Water Research X*, *3*, 100030.

Cózar, A., Echevarría, F., González-Gordillo, J. I., Irigoien, X., Úbeda, B., Hernández-León, S., ... & Fernández-de-Puelles, M. L. (2014). Plastic debris in the open ocean. *Proceedings of the National Academy of Sciences*, *111*(28), 10239–10244.

Crawford, C. B., & Quinn, B. (2016). *Microplastic Pollutants* (p. 336). Elsevier Science, Elsevier publications (ISBN: 9780128094068).

Driedger, A. G., Dürr, H. H., Mitchell, K., & Van Cappellen, P. (2015). Plastic debris in the Laurentian Great Lakes: A review. *Journal of Great Lakes Research*, *41*(1), 9–19.

Dris, R., Gasperi, J., Mirande, C., Mandin, C., Guerrouache, M., Langlois, V., & Tassin, B. (2017). A first overview of textile fibers, including microplastics, in indoor and outdoor environments. *Environmental Pollution*, *221*, 453–458.

Dris, R., Gasperi, J., Saad, M., Mirande, C., & Tassin, B. (2016). Synthetic fibers in atmospheric fallout: A source of microplastics in the environment? *Marine Pollution Bulletin*, *104*(1–2), 290–293.

Eriksen, M., Mason, S., Wilson, S., Box, C., Zellers, A., Edwards, W., ... & Amato, S. (2013). Microplastic pollution in the surface waters of the Laurentian Great Lakes. *Marine pollution bulletin*, *77*(1–2), 177–182.

Eriksen, M., Lebreton, L. C., Carson, H. S., Thiel, M., Moore, C. J., Borerro, J. C., ... & Reisser, J. (2014). Plastic pollution in the world's oceans: More than 5 trillion plastic pieces weighing over 250,000 tons afloat at sea. *PloS One*, *9*(12), e111913.

Farrell, P., & Nelson, K. (2013). Trophic level transfer of microplastic: Mytilus edulis (L.) to Carcinus maenas (L.). *Environmental Pollution*, *177*, 1–3.

Fendall, L. S., & Sewell, M. A. (2009). Contributing to marine pollution by washing your face: Microplastics in facial cleansers. *Marine Pollution Bulletin*, *58*(8), 1225–1228.

Free, C. M., Jensen, O. P., Mason, S. A., Eriksen, M., Williamson, N. J., & Boldgiv, B. (2014). High-levels of microplastic pollution in a large, remote, mountain lake. *Marine Pollution Bulletin*, *85*(1), 156–163.

Gewert, B., Plassmann, M. M., & MacLeod, M. (2015). Pathways for degradation of plastic polymers floating in the marine environment. *Environmental Science: Processes & Impacts*, *17*(9), 1513–1521.

Geyer, R., Jambeck, J. R., & Law, K. L. (2017). Production, use, and fate of all plastics ever made. *Science Advances*, *3*(7), e1700782.

GESAMP 2019, Guidelines for the Monitoring and Assessment of Plastic Litter in the Ocean.http://www.gesamp.org/publications/guidelines-for-the-monitoring-and -assessment-of-plastic-litter-in-the-ocean

He, P., Chen, L., Shao, L., Zhang, H., & Lü, F. (2019). Municipal solid waste (MSW) landfill: A source of microplastics?-Evidence of microplastics in landfill leachate. *Water Research*, *159*, 38–45.

Hidalgo-Ruz, V., Gutow, L., Thompson, R. C., & Thiel, M. (2012). Microplastics in the marine environment: A review of the methods used for identification and quantification. *Environmental Science & Technology*, *46*(6), 3060–3075.

Iñiguez, M. E., Conesa, J. A., & Fullana, A. (2017). Microplastics in Spanish table salt. *Scientific Reports*, *7*(1), 8620.

Jambeck, J. R., Geyer, R., Wilcox, C., Siegler, T. R., Perryman, M., Andrady, A., ... & Law, K. L. (2015). Plastic waste inputs from land into the ocean. *Science*, *347*(6223), 768–771.

Karami, A., Golieskardi, A., Choo, C. K., Larat, V., Galloway, T. S., & Salamatinia, B. (2017). The presence of microplastics in commercial salts from different countries. *Scientific Reports*, *7*, 46173.

Klein, S., Worch, E., & Knepper, T. P. (2015). Occurrence and spatial distribution of microplastics in river shore sediments of the Rhine-Main area in Germany. *Environmental Science & Technology*, *49*(10), 6070–6076.

Kooi, M., Reisser, J., Slat, B., Ferrari, F. F., Schmid, M. S., Cunsolo, S., ... & Schoeneich-Argent, R. I. (2016). The effect of particle properties on the depth profile of buoyant plastics in the ocean. *Scientific Reports*, *6*, 33882.

Lassen, C., Hansen, S. F., Magnusson, K., Hartmann, N. B., Jensen, P. R., Nielsen, T. G., & Brinch, A. (2015). Microplastics: Occurrence, effects and sources of releases to the environment in Denmark. Environmental project No. 1793,Danish Environmental Protection Agency. https://www.tilogaard.dk/Miljostyrelsens _rapport_om_mikroplast_978-87-93352-80-3.pdf

Law, K. L., & Thompson, R. C. (2014). Microplastics in the seas. *Science*, *345*(6193), 144–145.

Law, K. L., Morét-Ferguson, S. E., Goodwin, D. S., Zettler, E. R., DeForce, E., Kukulka, T., & Proskurowski, G. (2014). Distribution of surface plastic debris in the eastern Pacific Ocean from an 11-year data set. *Environmental Science & Technology*, *48*(9), 4732–4738.

Lebreton, L. C., Van der Zwet, J., Damsteeg, J. W., Slat, B., Andrady, A., & Reisser, J. (2017). River plastic emissions to the world's oceans. *Nature Communications*, *8*, 15611.

Lechner, A., Keckeis, H., Lumesberger-Loisl, F., Zens, B., Krusch, R., Tritthart, M., ... & Schludermann, E. (2014). The Danube so colourful: A potpourri of plastic litter outnumbers fish larvae in Europe's second largest river. *Environmental Pollution*, *188*, 177–181.

Magnusson, K., & Norén, F. (2014). Screening of microplastic particles in and downstream a wastewater treatment plant. http://naturvardsverket.diva-portal.org/ smash/get/diva2:773505/FULLTEXT01.pdf

McCormick, A., Hoellein, T. J., Mason, S. A., Schluep, J., & Kelly, J. J. (2014). Microplastic is an abundant and distinct microbial habitat in an urban river. *Environmental Science & Technology*, *48*(20), 11863–11871.

Murphy, F., Ewins, C., Carbonnier, F., & Quinn, B. (2016). Wastewater treatment works (WwTW) as a source of microplastics in the aquatic environment. *Environmental Science & Technology*, *50*(11), 5800–5808.

Murphy, F., Russell, M., Ewins, C., & Quinn, B. (2017). The uptake of macroplastic & microplastic by demersal & pelagic fish in the Northeast Atlantic around Scotland. *Marine Pollution Bulletin*, *122*(1–2), 353–359.

Peng, G., Zhu, B., Yang, D., Su, L., Shi, H., & Li, D. (2017). Microplastics in sediments of the Changjiang Estuary, China. *Environmental Pollution*, *225*, 283–290.

Plastics Europe (2019). Plastics – the Facts 2019 An analysis of European plastics production, demand and waste data, Published on the occasion of the special show of K 2019. A project of the German plastics industry under the leadership of Plastics Europe Deutschland e. V. and Messe Düsseldorf.

Ramasamy, E. V., Sruthi, S. N., Harit A. K., & Babu, N. (2019). Microplastics in human consumption: Table salt contaminated with microplastics. In S. Babel, A. Haarstrick, M. S. Babel, A. Sharp (Eds.), *Microplastics in the Water Environment* (pp. 74–80). Cuvillier Verlag. (ISBN 978-3-7369-7089-2 eISBN 978-3-7369-6089-3).

Setälä, O., Fleming-Lehtinen, V., & Lehtiniemi, M. (2014). Ingestion and transfer of microplastics in the planktonic food web. *Environmental Pollution*, *185*, 77–83.

Siegfried, M., Koelmans, A. A., Besseling, E., & Kroeze, C. (2017). Export of microplastics from land to sea. A modelling approach. *Water Research*, *127*, 249–257.

Sruthy, S and Ramasamy, E. V. (2017). Microplastic pollution in Vembanad Lake, Kerala, India: The first report of microplastics in lake and estuarine sediments in India. *Environmental Pollution* (Elsevier), *222*, 315–322.

Thompson, R. C., Olsen, Y., Mitchell, R. P., Davis, A., Rowland, S. J., John, A. W., ... & Russell, A. E. (2004). Lost at sea: where is all the plastic? *Science*, *304*(5672), 838–838.

UNEP (2017). *Microplastic an Emerging Issue*. United Nations Environment Programme.

UNEP (2018). *Single-Use Plastics: A Roadmap for Sustainability*. United Nations Environment Programme. 2018 ISBN: 978-92-807-3705-9 DTI/2179/JP.

Van Sebille, E., Wilcox, C., Lebreton, L., Maximenko, N., Hardesty, B. D., Van Franeker, J. A., ... & Law, K. L. (2015). A global inventory of small floating plastic debris. *Environmental Research Letters*, *10*(12), 124006.

Von Moos, N., Burkhardt-Holm, P., & Köhler, A. (2012). Uptake and effects of microplastics on cells and tissue of the blue mussel Mytilus edulis L. after an experimental exposure. *Environmental Science & Technology*, *46*(20), 11327–11335.

Wagner, M., Lambert, S., & Lambert, M. W. (2018). *Freshwater Microplastics*. Springer International Publishing.

Yonkos, L. T., Friedel, E. A., Perez-Reyes, A. C., Ghosal, S., & Arthur, C. D. (2014). Microplastics in four estuarine rivers in the Chesapeake Bay, USA. *Environmental Science & Technology*, *48*(24), 14195–14202.

Microplastics in Human Food Items

2

2.1 INTRODUCTION

Microplastic contamination has been well reported in many natural environments, including water (surface and water column) and sediments of lakes and rivers, the open oceans of remote locations including polar waters, polar ice, deep sea sediments and also in the air (Eriksen et al., 2014; Cozer et al., 2014; Van Cauwenberghe et al., 2013; Baldwin et al., 2016; Obbard et al., 2014; Dris et al., 2015; Ramasamy et al., 2021). The ubiquitous distribution of MPs in oceans and seas also hints that the products originating from these aquatic systems may also be loaded with MPs. Several studies have documented the presence of MPs in marine biotic products – mainly the seafood items – like clams, fishes and crabs (Van Cauwenberghe and Janssen, 2014; Li et al., 2015). Similarly, sea salt – the abiotic product from the sea – is also reported to have microplastics (Yang et al., 2015; Karami et al., 2016). Hence, consumption of seafood and salt could be a significant route of exposure to MPs in humans. Several toxicity studies have indicated the risks to human health when plastic particles are ingested. Despite the level of significance associated with MPs contamination in human consumable items, the studies pertaining to synthetic polymers in human consumables are few.

Globally, 6,452 studies on MPs have been published so far, as per a Web of Science search with the keyword 'microplastics' (October 13, 2021), while reports on MPs in human food and beverages are relatively few (Figure 2.1). For instance, studies on MPs in fishes are fewer with 1,288, while on mussels there exist 373 reports.

As far as India is concerned, the total number of studies published on MPs is around 194; on MPs contamination in fishes 59; in salt 13; in mussels 11; in drinking water five, while only one report exists on crabs and honey. This statistics reflects that those studies pertaining to MPs contamination in food and

DOI: 10.1201/9781003201755-2

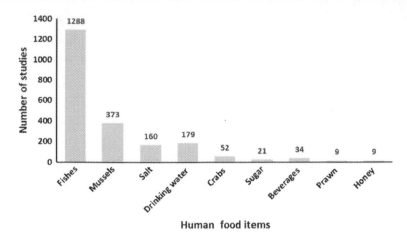

FIGURE 2.1   Microplastics contamination in human food items: global scenario. (Source: Web of Science.)

beverage items consumed by humans need more focus. This chapter reviews the status of global publications on MPs contamination in human food items.

2.2   MICROPLASTICS IN SEAFOOD ITEMS

The presence of MPs in various aquatic organisms (freshwater and marine) such as fishes, bivalves, shrimps, crabs, etc. has been well documented (Cole et al., 2013; Fossi et al., 2014; Cole et al., 2015; Rochman et al., 2015; Lusher et al., 2015; Ferreira et al., 2016; Ramasamy et al., 2019). Due to their tiny size MPs can be ingested by fishes, bivalves, shrimps, crabs, birds and other marine mammals (Rochman et al., 2015; Fossi et al., 2014) and easily introduced into the human food chain as most of them are favorite seafood items. A summary of the research findings of various studies on the occurrence of MPs in seafood items is given in Tables 2.1 and 2.2.

2.2.1   Microplastics in Fishes

Among the marine edible organisms, fishes are much studied and more studies have been reported on MPs contamination in fishes when compared with other organisms; maybe it is the seafood item most preferred by human beings.

TABLE 2.1 Microplastic Contamination in Edible Fishes

FISH SPECIES	ECOSYSTEM	COUNTRY	MICROPLASTICS		REFERENCE
			SHAPE	POLYMER TYPE	
Priacanthus hamrur Sciades sona Carangoides chrysophrys Benthopelagic species Harpadon nehereus Otolithoides pama Setipinna tenuifilis Coilia neglecta Anodontostoma chacunda Sardinella brachysoma Megalaspis cordyla	Marine	Bangladesh	Fibers, films, fragments, foams, granules	PE, PP, PS, PU, EPDM, SBR	Ghosh et al., 2021
Siganus canaliculatus Kuhila rupestris Valamugil speigleri	Marine Freshwater	Philippines	Fibers, fragments, foams, pellets, microbeads	–	Cabansag et al., 2021
Cyprinus carpio Pelteobagrus fulvidraco Mystus macropterus Pelteobagrus vachelli	Freshwater	China	Fibers, flakes, films, foams, lines, granules, strings	PE, PP, PB, PVC, PS, PET, PA, PES,	Zhang et al., 2021
Oreochromis niloticus Cirrhinus molitorella	Freshwater	China	Fibers, fragments, pellets	–	Sun et al., 2021
Calamus brachysomus Paralabrax maculatofasciatus Eucinostomus dowii Balistes polylepis Achirus mazatlanus Mugil curema	Marine	Mexico	Fibers	–	Jonathan et al., 2021

(Continued)

TABLE 2.1 (CONTINUED)   Microplastic Contamination in Edible Fishes

				MICROPLASTICS	
FISH SPECIES	ECOSYSTEM	COUNTRY	SHAPE	POLYMER TYPE	REFERENCE
Harpodon nehereus Chirocentrus dorab Sardinella albella Rastrelliger kanagurta Katsuwonus pelamis Istiophorus platypterus	Marine	India	Fibers	PE, PA, polyester	Sathish et al., 2020
Commercially available fishes	Marine	India	Fibers, films and pellets	PET, PA, PE	Karuppasamy et al., 2020
Sardinella longiceps Sardinella gibbose Stolephorus indicus Rastrelliger kanagurta Cyanoglossus macrostomus Thryssa mystax Lactarius lactarius Leiognathus splendens Nemipterus randalli Terapon puta Saurida tumbil Nibea maculata Epinephalus diacanthus Anodontostoma chacunda Dussumeria acuta Carangoides armatus	Marine	India	Fragments, filaments, pellets, fibers, films, foams	PE, PP, LDPE	James et al, 2020

(Continued)

TABLE 2.1 (CONTINUED) Microplastic Contamination in Edible Fishes

| FISH SPECIES | ECOSYSTEM | COUNTRY | MICROPLASTICS | | REFERENCE |
			SHAPE	POLYMER TYPE	
Finfish and shellfishes	Marine	India	Fibers	PVC, PU, EVAl, EVAc, acrylonitrile, nylon, polyisoprene, polyphenylene sulfide, acrylic (acryl fiber), poly etherimide, Surlyn ionomer	Goswami et al., 2020
Major consumed marine fishes Acanthopagrus latus Eleutheronemaa tetradactylum Lutjanus quinquelineatus	Marine	Kuwait	Fragments	PE, PO, PA	Al-Salem et al., 2020
Scomber spp. Truchurus truchurus Sardina pilchardus	Marine	Morocco	–	PA, PS, PC	Maaghloud et al., 2020
Hemiculter leucisculus	Freshwater	China	Fibers, fragments	–	Li et al., 2020

(Continued)

TABLE 2.1 (CONTINUED)   Microplastic Contamination in Edible Fishes

			MICROPLASTICS		
FISH SPECIES	ECOSYSTEM	COUNTRY	SHAPE	POLYMER TYPE	REFERENCE
Megalaspis cordyla Epinephelus coioides Rastrelliger kanagurta Euthynnus affinis Thunnus tonggol Eleutheronema tridactylum Clarias gariepinus Colossoma macropomum Nemipterus bipunctatus Ctenopharyngodon idella Selar boops	Marine	Malaysia	Fragments, fibers	PE, PP, PET	Karbalaei et al., 2019
Thryssa kammalensis Amblychaeturichthys hexanema Odontamblyopus rubicundus Cynoglossus semilaevis Chaeturichthys stigmatias Collichthys lucidus	Marine	China	Fibers, fragments, sheets	CP, PP, PE, PET	Feng et al., 2019

(Continued)

TABLE 2.1 (CONTINUED)   Microplastic Contamination in Edible Fishes

| FISH SPECIES | ECOSYSTEM | COUNTRY | MICROPLASTICS | | REFERENCE |
			SHAPE	POLYMER TYPE	
Cephalopholis urodeta Melichthys vidua Balistapus undulates Balistes capistratus Odonus niger Xanthichthys caeruleolineatus Pseudobalistes fuscus Acanthurus pyroferus Acanthurus japonicus Naso lituratus Naso brevirostris Acanthurus lineatus Pomacanthus imperator Dasyatis kuhlii Monocentrus japonica Zebrasoma veliferum	Marine	China	Fibers, films, fragments, pellets	PVC, PA, PE, PP	Nie et al., 2019
Leiognathus rivulatus Leiognathus equulus Monodactylus argeneteus Gerres oyena Herklotsichthys quadrimaculatus Atherinomorus lacunosus	Marine	Saudi Arabia	Fragments, granules, films, fibers	PS, PE, polyester, polyacryl amide, polyacrylic acid	Al-Lihaibi et al., 2019

(Continued)

TABLE 2.1 (CONTINUED)   Microplastic Contamination in Edible Fishes

| | | | MICROPLASTICS | | |
FISH SPECIES	ECOSYSTEM	COUNTRY	SHAPE	POLYMER TYPE	REFERENCE
Harpadon nehereus *Harpadon translucens* *Sardinella gibbose*	Marine	Bangladesh	Fragments, fibers, films	PET, PA	Hossain et al., 2019
Deep sea fishes	Marine	China	Films	–	Zhu et al., 2019
Commercial fishes	Marine	Chile	Microfibers	PE, PET	Pozo et al., 2019
Oreochromis niloticus	Freshwater	China	–	PS	Zhang et al., 2019
Lateolabrax maculatus	Estuary	China	–	–	Su et al., 2019
Sardine and anchovy	Estuary	France	Fibers	PET	Lefebvre et al., 2019
Prochilodus lineatus	Freshwater	South America	Fibers	–	Blettler *et al.,* 2019
Demersal fish	Freshwater	Hong Kong	Fragments, fibers	–	Chan et al., 2019
Danio rerio	Freshwater	–	Fibers, fragments	–	Qiao et al., 2019
Gambusia holbrooki	Freshwater	Australia	Fibers	Polyester	Su *et al.,* 2019
Carassius auratus	Freshwater	China	Fibers	PP, PE	Yuan et al., 2019
Eel, loach and crayfish	Rice–fish co-culture ecosystems	China	Fibers	PE, PP	Lv et al., 2019
Wild-caught fishes	Fishing ground	China	Fibers	Polyester	Zhang et al., 2019

(Continued)

TABLE 2.1 (CONTINUED) Microplastic Contamination in Edible Fishes

| | | | MICROPLASTICS | | |
| | | | | | |
FISH SPECIES	ECOSYSTEM	COUNTRY	SHAPE	POLYMER TYPE	REFERENCE
Thryssa kammalensis, Amblychaeturichthys hexanema, Odontamblyopus rubicundus, Cynoglossus semilaevis, Chaeturichthys stigmatias Collichthys lucidus	farm and marine (mariculture area)	China	–	PS, PVC	Fang et al., 2019
Gobio gobio	Freshwater	Belgium	Fibers, films, foams, fragments, pellets	EVA, PP, PET, PVC, CP, PVA, PA	Slootmaekers et al., 2019
Dicentrarchus labrax Diplodus vulgaris Platichthys flesus	Estuary	Portugal	Fibers	Polyester, PP, rayon	Bessa et al., 2018
Acanthopagrus australis Mugil cephalus	Estuary	Australia	Fibers	Polyester, acrylic-polyester blend, rayon	Halstead et al., 2018
Bigeye sculpin Triglops nybelini	Arctic ecosystem	Northeast Greenland	–	Polyester	Morgana et al., 2018
Sardine pilchardus Pagellus erythrinus Mullus barbatus	Marine	Mediterranean Sea	Fragments	PE	Digka et al., 2018
Sardine pilchardus Engraulis encrasicolus	Marine	Spain	Fibers	–	Compa et al., 2018

(Continued)

TABLE 2.1 (CONTINUED)   Microplastic Contamination in Edible Fishes

FISH SPECIES	ECOSYSTEM	COUNTRY	SHAPE	POLYMER TYPE	REFERENCE
				MICROPLASTICS	
Gonostoma denudatum
Serrivomer beanie
Lampanyctus macdonaldi	Marine	Canada	Fibers	PE	Wieczorek et al., 2018
Odontesthes regia
Strangomera bentincki
Sardinops sagax
Opisthonema libertate
Cetengraulis mysticetus
Engraulis ringens
Scomber japonicas	Marine	Southeast Pacific Ocean	Threads, fragments	P, PP	Ory et al., 2018
Rastrilleger kanagurta
Epinephalus menu	Freshwater	India	Fragments	PE, PP	Kumar et al., 2018
Rutilus rutilus (roach)	Estuary	United Kingdom	Fibers, fragments, films	PE, PP, polyester	Horton et al., 2018
Fishes	Freshwater	USA	Fibers	PE, PVC, PET, PS	Wagner et al., 2017
Hoplosternum littorale	Freshwater	Brazil	Fibers	–	Silva-Cavalcanti et al., 2017
Chelon subviridis
Johnius belangerii
Rastrelliger kanagurta
Stolephorus waitei	Dry fishes	Malaysia	Fragments, films, filaments	PE, PP, PS, PET, NY 6 phthaloc-yanine actinolite	Karami et al., 2017
Engraulis japonicas	Marine	Japan	Fragments, beads, microbeads	PP, PE	Tanaka & Takada, 2016

(Continued)

TABLE 2.1 (CONTINUED) Microplastic Contamination in Edible Fishes

FISH SPECIES	ECOSYSTEM	COUNTRY	MICROPLASTICS		REFERENCE
			SHAPE	POLYMER TYPE	
Marine and freshwater fishes	Marine Freshwater	China	Fibers, fragments, pellets, meso-fibers, meso-sheets	26 polymer types – CP, PET, polyester	Jabeen et al., 2017
Alosa fallax, Argyrosomus regius, Boops boops, Brama brama, Dentex macrophthalmus, Helicolenus dactylopterus, Lepidorhombusboscii, Lepidorhombus whiffiagonis, Lophius piscatorius, Merluccius merluccius, Merluccius merluccius, Mullussurmuletus, Mullus surmuletus, Pagellus acarne, Polyprion americanus, Raja asterias, Sardina pilchardus, Scomber japonicus, Scomber scombrus, Scyliorhinus canicula, Scyliorhinus canicula, Solea solea, Torpedo torpedo, Trachurus picturatus, Trachurus trachurus, Trichiurus lepturus, Trigla lyra, Trisopterus luscus, Zeus faber	Marine	Portugal	Fibers, fragments	PP, PE, PS, NY, rayon	Neves et al., 2015

(Continued)

TABLE 2.1 (CONTINUED) Microplastic Contamination in Edible Fishes

| | | | MICROPLASTICS | | |
FISH SPECIES	ECOSYSTEM	COUNTRY	SHAPE	POLYMER TYPE	REFERENCE
Fishes	Marine	United Kingdom	Fibers, fragments, beads	PA, PS, polyester, LDPE, acrylic,	Lusher et al., 2013

Notes:
PS: polystyrene; PE: polyethylene; PP: polypropylene; PVC: polyvinylchloride; PA: polyamide; PLA: polylactic acid; PVA: polyvinyl alcohol; HDPE: high-density polyethylene; LDPE: low-density polyethylene; CP: cellophane; PU: polyurethane; EVAl: ethylene-vinyl alcohol; EVAc: ethylene-vinyl acetate; EPDM: ethylene propylene diene monomer; SBR: styrene-butadiene rubber; PB: polybutylene; PES: polyethersulfone; PO: polyolefin; NY: nylon.

TABLE 2.2 Microplastic Contamination in Edible Bivalves (Mussels, Oysters and Clams)

SPECIES	MICROPLASTICS			REFERENCES
	SHAPE	SIZE	POLYMER TYPE	
Chlamys farreri Mytilus galloprovincialis Crassostrea gigas Ruditapes philippinarum	Fragments, fibers, films, granules	7–5,000 μm	PVC, PE, CPE, PET, PVDF, PVE, PEI, PVDC-PE, PVDC-PAN, CE, rayon	Ding et al., 2021
Oyster/mussel Manila clam	Fragments, fibers	50–5,000 μm	PS, PP, PA, PET, PE, polyester, acrylate polymer	Cho et al., 2021
Metapenaeus affinis	Fragments, films, fibers, spherules	100–1,000 μm	PP, PS, PET	Keshavarzifard et al., 2021
Perna viridis Meretrix meretrix	Micro-fragments	< 100 μm 21.06 μm–1.5 mm	PU, PVCA, plasticized PVC, PES, ABS, SBR, PET, PVK, PEVA	Dowarah et al., 2020
Carcinus maenas	Microspheres	8–10 μm	PS	Watts et al., 2014
Mytilus spp.	Microbeads	287μm – PE 204μm – PP	PE, PP	Revel et al., 2019
Mytilus edulis	Fragments	–	PLA, HDPE	Green et al., 2019
Mytilus edulis Crassostrea gigas Tapes philippinarum Patinopecten yessoensis	Fragments	300 μm	PE, PP, PS, polyester	Cho et al., 2019
Oyster	Fibers	< 1,500 μm	CP, PE, PET	Teng et al., 2019

(Continued)

TABLE 2.2 (CONTINUED)   Microplastic Contamination in Edible Bivalves (Mussels, Oysters and Clams)

SPECIES	MICROPLASTICS			REFERENCES
	SHAPE	SIZE	POLYMER TYPE	
Anodonta anatina	fibers	–	PE	Berglund et al., 2019
Mytilus galloprovincialis	Fragments	–	PE	Digka et al., 2018
Pema canaliculus	Fragments	–	PE	Webb et al., 2019
Mytilus edulis	Fibers	–	Semi-synthetic polystyrene, nylon	Scott et al., 2019
Mussel byssus	Beads, fragments, fibers	–	PS, PA, polyester	Li et al., 2019
Mytilus galloprovincialis	Fragments, fibers	20–40 μm	PE PP, PET, PS, PLY, PVC	Gomiero et al., 2019
Perna perna	Fibers, fragments	–	Nylon, PMMA	Birnstiel et al., 2019
Perna viridis	Fibers	5–25 μm	PS	Naidu, 2019
Perna viridis	Fibers, films, Particles	–	PS, polyketone, polybutadiene, phenol sulfon phthalein	Harit et al., 2019
Perna perna	Microbeads	0.1–1.0 μm	PVC	Santana et al., 2018
Mytilus galloprovincialis	Microbeads	3 μm	PS	Capolupo et al., 2018
Mytilus edulis	Fibers	–	Rayon, cotton fibers	Li et al., 2018
Mytilus galloprovincialis	Fragments	–	PE	Digka et al., 2018
Mytilus galloprovincialis	Microbeads	20–25 μm	LDPE	Pittura et al., 2018

(Continued)

TABLE 2.2 (CONTINUED) Microplastic Contamination in Edible Bivalves (Mussels, Oysters and Clams)

		MICROPLASTICS			
SPECIES	SHAPE	SIZE	POLYMER TYPE		REFERENCES
Mytilus galloprovincialis	Microbeads	1–50 μm	HDPE		Détrée, & Gallardo-Escárate, 2018
Dreissena polymorpha	Microbeads	1–10 μm	PS		Magni et al., 2018
Mytilus galloprovincialis	–	1,150–2,290 μm	–		Renzi, et al., 2018
Crassostrea virginica	Microbeads	50 nm–3 mm	PS		Gaspar et al., 2018
Pinctada margaritifera	Microbeads	6–10 μm	PS		Gardon et al., 2018
Abra nitida Ennucula tenuis	Microbeads	4–500 μm	PE		Bour et al., 2018
Corbicula fluminea	Microbeads	1–5 μm	–		Guilhermino et al., 2018
Mytilus galloprovincialis	Fibers, fragments, granules	0.66 ± 0.70 mm	CP, PP		Ding et al., 2018
Scrobicularia plana	Microbeads	20 μm	PS		Ribeiro et al., 2017
Atactodea striata	Microbeads	63–250 μm	PS		Xu et al., 2017
Mytilus edulis Ostrea edulis	Microbeads	65.6–102.6 μm	PLA, HDPE		Green et al., 2017, 2016
Mytilus spp.	Microbeads	2–6 μm	PS		Paul-Pont et al., 2016
Perna viridis	Microbeads	1–50 μm	PVC		Rist et al., 2016
Crassostrea gigas	Microbeads	2–6 μm	PS		Sussarellu et al., 2016

(Continued)

TABLE 2.2 (CONTINUED) Microplastic Contamination in Edible Bivalves (Mussels, Oysters and Clams)

SPECIES	MICROPLASTICS			REFERENCES
	SHAPE	SIZE	POLYMER TYPE	
Commercial bivalves from a fishery market	Fibers, fragments, pellets	250 μm	–	Li et al., 2015
Mytilus edulis	Microbeads	10–90 μm	PS	Van Cauwenberghe, et al., 2015
Mytilus galloprovincialis	–	< 100 μm	PE, PS	Avio et al., 2015
Mytilus galloprovincialis	Microbeads	50 nm	PS-NH$_2$	Canesi et al., 2015
Mytilus edulis	Microbeads	0–80 μm	HDPE	Von Moos et al., 2012
Mytilus edulis	Microbeads	30 nm	PS	Wegner et al., 2012

Notes:
PS: polystyrene; PE: polyethylene; PP: polypropylene; PVC: polyvinylchloride; PLA: polylactic acid; HDPE: high-density polyethylene; LDPE: low-density polyethylene; CP: cellophane; PMMA: polymethyl methacrylate; PU: polyester urethane; PVCA: vinyl chloride/vinyl acetate copolymer; PES: polyp-phenylene ether sulfone); PET: polyethylene terephthalate; ABS: acrylonitrile/butadiene/styrene copolymer; SBR: butadiene/styrene copolymer; PVK: poly(N-vinyl carbazole); PEVA: PVC/acrylic alloy and poly(ethylene-co-vinyl acetate); CPE: chlorinated polyethylene; PVDF: polyvinylidene fluoride; PVDC-PE: polyvinylidene chloride–polyethylene; PVF: polyvinyl ester; PEI: polyetherimide; PVDC-PAN: polyvinylidene chloride–polyacrylonitrile; PS-NH$_2$: polystyrene nanoparticles.

From various studies reported by researchers (Li et al., 2020; Sathish et al., 2020; Maaghloud et al., 2020; James et al., 2020; Goswami et al., 2020; Al-Salem et al., 2020; Fang et al., 2019; Yin et al., 2019; Chan et al., 2019; Blettler et al., 2019; Garnier et al., 2019; Compa et al., 2018; Horton et al., 2018; Kumar et al., 2018; Digka et al., 2018; Baalkhuyur et al., 2020; Bessa et al., 2018; Wang et al., 2017; Silva-Cavalcanti et al., 2017; Lusher, 2015; Choy and Drazen, 2013; Davison and Asch, 2011; Boerger et al., 2010) it is apparent that microplastics (mainly fibers, fragments and films) are ingested by the fishes; among the different shapes of MPs, fiber-shaped MP particles are predominantly found in fishes (Table 2.1). Globally, China (235 reports) is leading with a greater number of studies on MPs contamination in fishes, followed by the USA, while India stands in 12th position (Figure 2.2) with 59 publications. Out of these 59 papers from India, ten are review articles and 49 are research articles. After careful screening of these 49 publications, we could find only a few publications (Table 2.1) dealing with MPs contamination in edible fishes.

Among the different aquatic systems, a greater number of studies have been reported on marine fishes than estuarine and freshwater fishes (Table 2.1). Morgana et al. (2018) reported MPs contamination in Arctic fishes (*Triglops nybelini* and *Boreogadus saida*) from northeast Greenland in the Arctic. Their findings state that more than one plastic particle has been found in the digestive tract of the fishes and fibers were the dominant shape of particles. The MPs extracted from Arctic fishes were composed of polyester, acrylic, polyamide, polyethylene and ethylene-vinyl acetate; among these, polyester was the dominant polymer (Morgana et al., 2018). Lusher et al. (2013) documented MPs contamination in ten species of fishes from the English Channel; polyamide and rayon were identified as the most common types of polymers. However, many researchers have reported polyethylene (PE) as the dominant type of polymer in marine fishes, followed by polyethylene terephthalate (PET) and polypropylene (PP). Studies on freshwater fishes confirm polypropylene as the predominant polymer while polyester was common in estuarine fishes (Table 2.1).

MPs were observed in the stomach contents of three small pelagic species in the central zone of the Atlantic, located between Cape Cantin (33°N) and Cape Boujdor (26°N): *Scomber* spp., *Truchurus truchurus* and *Sardina pilchardus* (Maaghloud et al., 2020). The presence of three polymers – polyamide (PA), acrylic (PC) and polystyrene (PS) – in 26% of the individuals was reported in this study. Al-Salem et al. (2020) reported MPs in the gut content of marine fish *Acanthopagrus latus*, *Eleutheronemaa tetradactylum* and *Lutjanus quinquelineatus* in Kuwait. Microscopic and spectroscopic analysis indicated polyethylene (PE) as the polymer component and the MPs were noted to have yellowish color and fragments in shape.

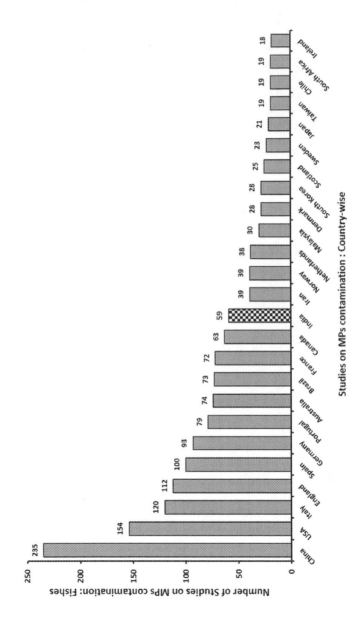

FIGURE 2.2 Studies on microplastic contamination in fishes: global scenario. (Source: Web of Science.)

The earliest publication from India on MPs-contaminated fishes (Kumar et al., 2018) has reported MPs contamination in two species of fishes from India (*Rastrilleger kanagurta* and *Epinephalus merra*) sampled from a fish landing site at Tuticorin, Tamil Nadu, India. They have reported fiber-shaped MPs as the predominant one, followed by fragments of black, red and translucent colors. Polyethylene and polypropylene polymers were the polymer type of MPs found in both species of fishes.

Sathish et al. (2020), recently reported MPs contamination in the gastrointestinal tract of six fishes (*Harpodon nehereus, Chirocentrus dorab, Sardinella albella, Rastrelliger kanagurta, Katsuwonus pelamis* and *Istiophorus platypterus*), collected from Tuticorin on the southeast coast of India. Most of the MPs identified were of blue color, of fiber type and their size was 500 µm. Polyethylene was the most commonly detected polymer, followed by polyester and polyamide.

Karuppasamy et al. (2020) found MPs in commercially available fishes in Chennai and Nagapattinam in Tamil Nadu, on the southeast coast of India. Polymers found in gastrointestinal tracts were polyethylene, polyamide and polyester. Goswami et al. (2020) detected 11 different types of MPs in the gastrointestinal tract of finfish and shellfishes collected from the Port Blair Bay, Andaman Islands; among these, Surlyn ionomer, acryl and nylon fibers were the most abundant polymers. The size of MPs was ranging from 111.58 to 5,094 µm with an average of 1,308.46 ± 1,016.73 µm.

A recent study conducted on the gut content of 16 species (653 individuals) comprising pelagic (eight species) and demersal (eight species) fishes indicated the occurrence of microplastics (fragment > filament > pellet) of size 0.27 mm to 3.2 mm in *Sardinella longiceps, S. gibbosa, Stolephorus indicus, Rastrelliger kanagurta* and *Cyanoglossus macrostomus*. Raman spectroscopy indicated that polyethylene (PE) and polypropylene (PP) were the polymer types (James et al., 2020).

The research publications reviewed in this chapter suggest direct ingestion as the prime route of MPs found in the fishes, either by ingesting the target prey which was already contaminated with MPs or by ingesting the tiny microplastic particles mistaking them for prey items. Studies also mention false satiation and reduced feed intake as the consequences of MPs ingestion but more studies are needed in order to establish such conclusions. Also, there are reports (Santos and Jobling, 1992) on the Atlantic cod (*Gadus morhua*), a benthopelagic fish, stating that MPs (beads of 2 mm size) were excreted upon ingestion, while beads of 5 mm size were held for a longer period of time. Such studies are lacking in other common edible fishes and such studies will provide significant information that smaller MPs have lesser implications than larger ones as far as the food chain/web transfer and trophic level transfer of MPs are concerned.

2.2.2 Microplastics in Bivalves

Bivalves are the most highly consumed seafood item next to fishes. Several studies have been conducted globally on MPs contamination in bivalves (Dowarah et al., 2020; Harit et al., 2019; Berglund et al., 2019; Brinstiel et al., 2019; Gomiero et al., 2019; Webb et al., 2019; Scott et al., 2019; Li et al., 2019, 2018; Naidu, 2019; Graham et al., 2019; Cho et al., 2019; Revel et al., 2019; Teng et al., 2019; Magni et al, 2018; Pittura et al., 2018; Digka et al., 2018; Naji et al., 2018; Renzi et al., 2018; Santana et al., 2018; Green et al., 2018, 2016; Gaspar et al., 2018; Guilhermino et al., 2018; Ribeiro et al., 2017; Sussarellu et al., 2016; Rist et al., 2016; Van Cauwenberghe et al., 2015; Avio et al., 2015; Canesi et al., 2015; Wegner et al., 2012). From these studies, it is noted that MPs in the form of microbeads, fragments, fibers and pellets have been found in the soft tissue, digestive tract, valves and gills of mussels, oysters and clams. Li et al. (2018) report very specifically about 80% of MPs are fibers in shape in mussels collected from the coastal waters of China; Davidson and Dudas (2016) report that 90% of MPs are fibers in Manila clams (*Venerupis philippinarum*) from Baynes Sound, British Columbia. A wide range of polymer profiles of MPs found in bivalves has been reported in various studies, viz. polystyrene (PS), polyethylene (PE), polypropylene (PP), polyethylene terephthalate (PET) polyvinylchloride (PVC), polyacrylonitrile (PAN), high-density polyethylene (HDPE), low-density polyethylene (LDPE), polylactic acid (PLA), cellophane (CP) and nylon (Table 2.2). Of these, PS was the most dominantly found polymer among all the studies in bivalves, followed by PP and PE.

A country-wise survey of studies pertaining to MPs contamination in bivalves reveals that China is the leading nation with 52 studies, followed by Italy and France (Figure 2.3). Only nine studies are reported from India. Naidu (2019) has reported the presence of PS polymers in MPs extracted from green mussels, *Perna viridis*, collected from the fishing harbor of Chennai on the southeast coast of India. Harit et al. (2019) extracted MPs from green mussels (*Perna viridis*), locally known as 'Kallumakkaya', collected from coastal regions of Thiruvananthapuram, Kozhikode and Kannur of Kerala, India. The polymer profile of the MPs reported in this study includes polystyrene, polyketone and polybutadiene. A study by Dowarah et al. (2020) on *Perna viridis* and *Meretrix meretrix* from three estuaries (Ariyankuppam, Panithittu and Chunnambar) of Puducherry, India found 12 types of polymers (polyester urethane [PU], plasticized PVC, poly[p-phenylene ether sulfone] [PES], vinyl chloride/vinyl acetate copolymer [PVCA], polyethylene terephthalate [PET], acrylonitrile/butadiene/styrene copolymer [ABS], butadiene/styrene copolymer [SBR], vinyl chloride/vinyl acetate copolymer [PVCA], poly[N-vinyl carbazole] [PVK], PVC/acrylic alloy, poly[ethylene-co-vinyl acetate] [PEVA] and polyethylene terephthalate [PET]) of micro-fragment shapes. Qu et al.

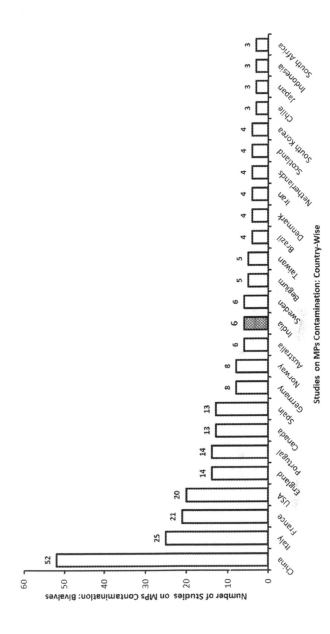

FIGURE 2.3   Studies reported on microplastic contamination in bivalves: global scenario. (Source: Web of Science.)

(2018) reported the significant correlations between MPs in mussels and their surrounding waters.

2.2.3 MPs in Crustaceans: Crabs, Prawns and Shrimps

Ingestion of MPs contamination in crustaceans has been poorly documented compared to fishes and bivalves. Globally, 76 studies found MPs in crustaceans as per the Web of Science database with the keywords "microplastics + crustaceans". Of these, only three research articles (Mishra et al., 2019; Daniel et al., 2020; Thiagarajan et al., 2021) and one review (Ajith et al. 2020) are reported from India. Mishra et al. (2019) assessed the toxic behavior of nanoplastics (polystyrene nano-spheres [PNS]) toward blood cells and marine crustaceans. Meanwhile, Daniel et al. (2020) studied the abundance and seasonal variation of MPs in *Fenneropenaeus indicus* (Indian white shrimps). Forbes (2019) found MPs in fiddler crabs (genus: *Uca*) and reported the dominant shape of MPs in fiddler crabs as microfibers. Horn et al. (2019) reported Pacific mole crabs found on California beaches have ingested microplastics. The occurrence of MPs has also been reported in Antarctic krill (*Euphausia superba*) in which microplastics were noted with a size ranging from 31.5 μm to < 1 μm (Dawson et al., 2018). Abbasi et al. (2018) found fibrous fragment MP particles in the exoskeleton and muscle of the tiger prawn (*Penaeus semisulcatus*). MPs were also recorded in the gastrointestinal tract of brown shrimps (*Metapenaeus monocerous*) and tiger shrimps (*Penaeus monodon*), collected from the northern Bay of Bengal, Bangladesh. A total of 33 and 39 MP particles were found in brown shrimps and tiger shrimps respectively. The dominant shape and color of MP particles were filaments and black. Polyamide and rayon were the polymers observed in the MPs (Hossain et al., 2020). Synthetic fibers ranging from 200 μm up to 1,000 μm in size were detected in brown shrimps (*Crangon crangon*; Linnaeus, 1758) from shallow water habitats of channel areas and coastal waters of the southern North Sea (Devriese et al., 2015).

2.3 MICROPLASTICS IN SALT, SUGAR AND HONEY

The ubiquitous distribution of MPs in seas and oceans and the consequent occurrence of MPs in seafood items have been well documented in the literature. Like seafood items, sea salt is also a common product (abiotic sea product) originating from seawater; hence, as expected, it is also reported to have MPs contamination (Yang et al., 2015; Iñiguez et al., 2017; Karami et al. 2017a; Seth and Shriwastav, 2018; Ramasamy et al., 2019; Feng et al., 2021; Sivagami et al., 2021; Yaranal et al., 2021). Sodium, an essential element in maintaining homeostasis in the human body, is obtained by consuming common salt (Brown et al., 2009). Commercial sea salts are produced by a crystallization process in which seawater on a large scale is subjected to evaporation using natural sunlight heat and wind. During this process, the possibility of transfer of seawater contaminants like microplastics into the sea salt is very high (Karamai et al., 2017) and thereby providing direct means of human ingestion (Seth and Shriwastav, 2018).

India is a salt-exporting country and it is the third largest producer of salt after China and the United States of America (USA); it exports edible and industrial salts to many countries (GoI, 2017). Three-fourths of the salt produced in India in 2015 was from seawater (Seth and Shriwastav, 2018). However, the studies reporting MPs contamination in sea salt from India are scarce. After global attention to the presence of MPs in table salt began in 2015 (Yang et al., 2015), currently there are about 160 research publications available globally on table salt contaminated with MPs.

To date only 13 studies from India have been published on Indian salts contaminated with MP (Seth and Shriwastav, 2018; Ramasamy et al., 2019; Yaranal et al., 2021; Vidyasakar et al., 2021; Nithin et al., 2021). The average concentration of MPs (34 particles/kg of salt) from six commercially available brands of sea salt has been reported by Ramasamy et al. (2019), while Seth and Shriwastav (2018) reported for eight brands of Indian salts. The observation of MPs contamination in Indian salts by Ramasamy et al. (2019) is quantitatively lesser than that of Seth and Shriwastav's observation; however, it is closer to what Gündoğdu et al. (2018) reported for Turkish sea salt and what was reported for Spanish salt by Iñiguez et al. (2017). Six types of polymers – polystyrene (PS), polyamide (PA), polyethylene terephthalate (PET), polypropylene (PP), polyethylene (PE) and polyvinylchloride (PVC) – were identified from the MP particles extracted from the commercial Indian seasalts, and polystyrene was the predominant polymer observed (Ramasamy et al., 2019). Meanwhile, Seth and Shriwastav (2018) reported four types of polymers: polyesters, polystyrene, polyamide and polyethylene.

When global studies on MPs contamination of salt are screened, around 27 polymer types have been detected from the different types and brands of salt samples, with dominant polymers of polyethylene terephthalate (PET), polyethylene (PE) and polypropylene (PP) as reported by Lee et al. (2019).

Karami et al. (2017) reported MPs contamination in 17 commercial brands of salts from eight different countries (Australia, France, Iran, Japan,

Malaysia, New Zealand, Portugal and South Africa). The common polymers observed by them in all these salts were polypropylene and polyethylene. Fragments followed by filaments were the predominant shape of MPs. Iñiguez et al. (2017) studied Spanish table salt (sea salt and well salt) and detected 50–280 MPs particles/kg of sea salt while the well salt had 115–185 particles/kg. Similarly, Gündoğdu (2018) and Yang et al. (2015) also reported MPs from different varieties of salt such as table salt (sea salt), lake salt, rock and well salt. Yang et al. (2015) reported a higher number of MPs in all type of salts (sea salt: 550–681 particles/kg; lake salt: 43–364 particles/kg; and rock/well salt: 7–204 particles/kg) compared to Gündoğdu (2018) (sea salt: 16–84 particles/kg; lake salt: 8–102 particles/kg; and rock salt: 9–16 particles/kg). However, from most of the studies, it can be noted that the highest numbers of MPs are found in sea salt, followed by lake and rock salts. Iñiguez et al. (2017) in their study conclude that the background level of MPs in the environment plays a significant role in the MP contamination of salt. However, they have ruled out the possibility of packaging material used in packing the salt as a source of MPs.

The average daily intake of salts for humans has been recommended by the WHO not to exceed 5 g (Iñiguez et al., 2017). Accordingly, for Indian salt, Seth and Srivastawa (2018) have given the MPs content as 103 particles per kg and Ramasamy et al. (2019) have noted 118 particles per kg as the maximum value they have obtained for commercial salt; when calculated for the per-day intake, as per WHO guidelines per annum, it becomes 187 MPs and 215 MPs respectively as per capita consumption of microplastic particles through Indian salt. Similar calculations for Spanish salts indicate 510 particles per capita per year (Iñiguez et al., 2017).

The presence of MPs in the sea salt is considered hazardous due to the polymer components of the MPs and the additives such as dyes, plasticizers, etc. added to it while manufacturing the plastics for various applications. Apart from these, when the MPs remain for years in seawater they adsorb many toxic chemicals including POPs, heavy metals and pesticides; after ingestion, these chemicals may be desorbed in the digestive tract of the organism and affect the health of the organism, or these chemicals may also enter the next higher level in the food chain/web. It is also reported that MPs also act as vectors of transporting many microbes which adhere and develop as biofilm over the surface of the MPs, causing harm to the concerned organism which ingests them (Ramasamy et al., 2019; Iñiguez et al., 2017).

The overall findings of the global studies reveal the presence of MPs in most of the brands of sea salt commercially available around the globe. Even though the number of MPs found in the Indian salt is lower than the MPs concentration reported in the rest of the world, the harmful impacts on human health cannot be ignored. Necessary steps need to be initiated to reduce the

TABLE 2.3 Microplastic Contamination in Salt, Sugar and Honey

MPS IN HUMAN FOOD ITEMS	COUNTRY	MICROPLASTICS			REFERENCES
		SHAPE	SIZE	POLYMER TYPE	
Sea salt crude salt, saturated brine, coarse salt, refined salt	China	Fibers, films, spheres, debris	18–4,936 μm	PE, PS, PET	Feng et al., 2021
	Nigeria Zimbabwe Malawi Cameroun Kenya Ghana Uganda South Africa	Fragments, fibers, granules	3.3–4,660 μm	PE, PP, PET, PEI, PVA	Fadare et al., 2021
Commercial salt – ten brands	India	Fragments, fibers, pellets	3.8 μm–5.2 mm	CP, PS, PA, PAE	Sivagami et al., 2021
Edible sea salt	India	Fibers, films, pellets	100–1,000 μm	PS, PVC LDPE, HDPE	Vidyasakar et al., 2021
Sea salt and bore-well salt	India	Fibers, fragments	55 μm–2 mm	PP, PES, PA, PE	Sathish et al., 2020
Sea salt	India	Fibers	–	PE, PP, CL, NY	Selvam et al., 2020
Table salt	India	Fibers, fragments, beads, sheets	500–1,000 μm	PS, PA, PET, PVC, PP, PE	Ramasamy et al., 2019

(Continued)

TABLE 2.3 (CONTINUED) Microplastic Contamination in Salt, Sugar and Honey

MPS IN HUMAN FOOD ITEMS	COUNTRY	MICROPLASTICS			
		SHAPE	SIZE	POLYMER TYPE	REFERENCES
Table salts (products)	Taiwan	Fragments, fibers	80–1,500 µm	PE, PEI, PES, PET, PAC, PP, PS	Lee et al., 2019
Sea, lake and rock salts	Turkey	Fibers, fragments, films	20–5,000 µm	PP, PE, PU, PET, PVC, PA-6	Gündoğdu 2018
Sea, lake and rock salts	USA Mexico Celtic Sea Atlantic Ocean Pacific Ocean Mediterranean Sea Sicilian Sea Himalayas USA	Fibers, fragments	40–5,000 µm	–	Kosuth et al., 2018
Table salts	Italy Croatia	Fibers, fragments, granules, films, foams	4–2,100 µm 15–4,628 µm	PP	Renzi and Blašković 2018
Sea salt	India	Fibers, fragments	500–2,000 µm	PS, PES, PA, PE, PET	Seth and Shriwastav 2018

(Continued)

TABLE 2.3 (CONTINUED)   Microplastic Contamination in Salt, Sugar and Honey

MPS IN HUMAN FOOD ITEMS	COUNTRY	MICROPLASTICS			REFERENCES
		SHAPE	SIZE	POLYMER TYPE	
Sea, lake and rock salts	Brazil	Fragments, fibers, sheets	100–1,000 µm	PET, PP	Kim et al., 2018
	Senegal		100–3,000 µm	PET, PP, PE	
	UK		100–2,000 µm	PVC, PE, PP	
	USA		100–1,000 µm	PP	
	Bulgaria		100–1,000 µm	NL, PP, PE, PVC	
	India		100–5,000 µm	NL, PP, PE, PVC, PET	
	Croatia		100–5,000 µm	NL, PP, PE, PET	
	Italy		100–3,000 µm	NL, PP, PE, PVC, PET, acrylic	
	Australia		100–4,000 µm		
	China		100–5,000 µm	NL, PE, PP, PU, PVC, EVA, PET	
	Chinese Taiwan		100–2,000 µm		
	Indonesia		100–3,000 µm	NL, PE, PP, PW, PVC, EVA, PET, acrylic	
	Korea		100–5,000 µm	PE, PET, PP	
	Thailand		100–5,000 µm		
	Vietnam		100–5,000 µm	Acrylic, PE, PP, PET, NL	
	Belarus		0–5,000 µm	PP, PE, PET, PVC	
	China		100–4,000 µm	PP, PE, PW, acrylic	
	Germany		100–2,000 µm	PET	
	Hungary		100–1,000 µm	Teflon, PP, PET,	
	Italy		100–1,000 µm	PET	
	Pakistan		100–2,000 µm	NL, PR, PVC, PET	
	Philippines		100–4,000 µm	PP, PE, PVC, PET	
	China			PP, PE, PET, EVA	
	Senegal			PP, PE, PET, PVC	
				PP, PS, PE, PET, Teflon	
				PE, PS, PET, PVC, acrylic	

(Continued)

TABLE 2.3 (CONTINUED) Microplastic Contamination in Salt, Sugar and Honey

MPS IN HUMAN FOOD ITEMS	COUNTRY	MICROPLASTICS				REFERENCES
		SHAPE	SIZE	POLYMER TYPE		
Table salts	China	Fragments	300 μm	PET		Zhang et al., 2018
Sea salts, well salts	Spain	Fibers	10–3,500 μm	PP, PE, PET		Iñiguez et al., 2017
Sea salts, lake salt	Australia France Japan New Zealand Portugal South Africa Malaysia Iran	Fragments, filaments, films	160–1,000 μm	PP, PE, PET, PS, NY6, polyacrylonitrile		Karami et al., 2017
Sea salts	China	Fragments, fibers, pellets, sheets	45–4,300 μm	PET, PE, CL		Yang et al. 2015
Honey	Ecuador	Fibers, fragments	5.15–3,302.68 μm	PP, HDPE, LDPE, PAAm		Diaz-Basantes et al., 2020
Honey	Switzerland	Fibers, particles	< 500 μm 10–20 μm	PET, CL		Mühlschlegel et al., 2017

(Continued)

TABLE 2.3 (CONTINUED) Microplastic Contamination in Salt, Sugar and Honey

MPS IN HUMAN FOOD ITEMS	COUNTRY	SHAPE	MICROPLASTICS SIZE	POLYMER TYPE	REFERENCES
Honey	Germany France Italy Spain Mexico	Fibers, fragments	40 μm–9 mm 10–20 μm	–	Liebezeit and Liebezeit, 2013
Sugar	Germany	Fibers, fragments	40 μm–6 mm 10–20 μm	–	Liebezeit and Liebezeit, 2013

Notes:
CP: cellophane; CL: cellulose; NY: nylon; PES: polyester; PA: polyamide; PP: polypropylene; PE: polyethylene; PET: polyethylene terephthalate; PEI: polyetherimide; PVA: polyvinyl alcohol; PAC: polyacetal or polyoxymethylene; PW: paraffin wax; PU: polyurethane; EVA: ethylene-vinyl acetate; PC: polycarbonate; PR: phenoxy resin; PAE: polyarylether; PAAm: polyacrylamide.

background concentration of MPs in the environment, i.e., in the seawater by adopting appropriate filtering techniques prior to pumping it into the salt pan for crystallization; also, steps are to be introduced to avoid airborne sources of MPs contamination in the production and packaging units of salt industries (Ramasamy et al., 2019).

Studies pertaining to contamination of MPs in sugar and honey are limited. Liebezeit and Liebezeit (2013) have observed the presence of MPs in honey sampled from Germany, France, Italy, Spain and Mexico. They have reported fiber-shaped MPs having a length ranging from 40 μm to about 9 mm and fragment-shaped MPs, having a size of about 10–20 μm and most of the fragments were found to be blue in color. The count of fibers ranged from 20 to 330 (mean value of 87 ± 73/500 g) per 500 g of honey, while that of fragments was relatively less (0–19/500 g; mean 4 ± 4/500 g of honey). Mühlschlegel et al. (2017) studied honey from Switzerland and reported fiber-shaped MPs (< 500 μm) in the honey; the polymer identified was PET (polyethylene terephthalate).

In the case of sugar, Liebezeit and Liebezeit (2013) studied five samples of commercially available sugars and found MPs of fibers and fragments. The length of fibers ranged from about 40 μm to about 6 mm, while fragments had 10–20 μm size. The number of fibers was 217 ± 123 fibers/kg (mean) and the number of fragments was 32 ± 7/kg (mean). The gist of various reports on microplastic contamination in salts, sugar and honey is given in Table 2.3.

REFERENCES

Abbasi, S., Soltani, N., Keshavarzi, B., Moore, F., Turner, A., & Hassanaghaei, M. (2018). Microplastics in different tissues of fish and prawn from the Musa Estuary, Persian Gulf. *Chemosphere*, *205*, 80–87.

Ajith, N., Arumugam, S., Parthasarathy, S., Manupoori, S., & Janakiraman, S. (2020). Global distribution of microplastics and its impact on marine environment: a review. *Environmental Science and Pollution Research*, *27*(21), 25970–25986.

Al-Lihaibi, S., Al-Mehmadi, A., Alarif, W. M., Bawakid, N. O., Kallenborn, R., & Ali, A. M. (2019). Microplastics in sediments and fish from the Red Sea coast at Jeddah (Saudi Arabia). *Environmental Chemistry*, *16*(8), 641–650.

Al-Salem, S. M., Uddin, S., & Lyons, B. (2020). Evidence of microplastics (MP) in gut content of major consumed marine fish species in the State of Kuwait (of the Arabian/Persian Gulf). *Marine Pollution Bulletin*, *154*, 111052.

Avio, C. G., Gorbi, S., Milan, M., Benedetti, M., Fattorini, D., d'Errico, G., ... & Regoli, F. (2015). Pollutants bioavailability and toxicological risk from microplastics to marine mussels. *Environmental Pollution*, *198*, 211–222.

Baalkhuyur, F. M., Qurban, M. A., Panickan, P., & Duarte, C. M. (2020). Microplastics in fishes of commercial and ecological importance from the Western Arabian Gulf. *Marine Pollution Bulletin*, *152*, 110920.

Baldwin, A. K., Corsi, S. R., & Mason, S. A. (2016). Plastic debris in 29 Great Lakes tributaries: Relations to watershed attributes and hydrology. *Environmental Science and Technology, 50*(19), 10377–10385.

Berglund, E., Fogelberg, V., Nilsson, P. A., & Hollander, J. (2019). Microplastics in a freshwater mussel (Anodonta anatina) in Northern Europe. *Science of the Total Environment, 697*, 134192.

Bessa, F., Barría, P., Neto, J. M., Frias, J. P., Otero, V., Sobral, P., & Marques, J. C. (2018). Occurrence of microplastics in commercial fish from a natural estuarine environment. *Marine Pollution Bulletin, 128*, 575–584.

Birnstiel, S., Soares-Gomes, A., & da Gama, B. A. (2019). Depuration reduces microplastic content in wild and farmed mussels. *Marine Pollution Bulletin, 140*, 241–247.

Boerger, C. M., Lattin, G. L., Moore, S. L., & Moore, C. J. (2010). Plastic ingestion by planktivorous fishes in the North Pacific Central gyre. *Marine Pollution Bulletin, 60*(12), 2275–2278.

Blettler, M. C., Garello, N., Ginon, L., Abrial, E., Espinola, L. A., & Wantzen, K. M. (2019). Massive plastic pollution in a mega-river of a developing country: Sediment deposition and ingestion by fish (Prochilodus lineatus). *Environmental Pollution, 255*, 113348.

Bour, A., Haarr, A., Keiter, S., & Hylland, K. (2018). Environmentally relevant microplastic exposure affects sediment-dwelling bivalves. *Environmental Pollution, 236*, 652–660.

Brown, B. B., Yamada, I., Smith, K. R., Zick, C. D., Kowaleski-Jones, L., & Fan, J. X. (2009). Mixed land use and walkability: Variations in land use measures and relationships with BMI, overweight, and obesity. *Health & Place, 15*(4), 1130–1141.

Cabansag, J. B. P., Olimberio, R. B., & Villanobos, Z. M. T. (2021). Microplastics in some fish species and their environs in Eastern Visayas, Philippines. *Marine Pollution Bulletin, 167*, 112312.

Canesi, L., Ciacci, C., Bergami, E., Monopoli, M. P., Dawson, K. A., Papa, S., ... & Corsi, I. (2015). Evidence for immunomodulation and apoptotic processes induced by cationic polystyrene nanoparticles in the hemocytes of the marine bivalve Mytilus. *Marine Environmental Research, 111*, 34–40.

Capolupo, M., Franzellitti, S., Valbonesi, P., Lanzas, C. S., & Fabbri, E. (2018). Uptake and transcriptional effects of polystyrene microplastics in larval stages of the Mediterranean mussel Mytilus galloprovincialis. *Environmental Pollution, 241*, 1038–1047.

Catarino, A. I., Macchia, V., Sanderson, W. G., Thompson, R. C., & Henry, T. B. (2018). Low levels of microplastics (MP) in wild mussels indicate that MP ingestion by humans is minimal compared to exposure via household fibres fallout during a meal. *Environmental Pollution, 237*, 675–684.

Chan, H. S. H., Dingle, C., & Not, C. (2019). Evidence for non-selective ingestion of microplastic in demersal fish. *Marine Pollution Bulletin, 149*, 110523.

Cho, Y., Shim, W. J., Jang, M., Han, G. M., & Hong, S. H. (2019). Abundance and characteristics of microplastics in market bivalves from South Korea. *Environmental Pollution, 245*, 1107–1116.

Cho, Y., Shim, W. J., Jang, M., Han, G. M., & Hong, S. H. (2021). Nationwide monitoring of microplastics in bivalves from the coastal environment of Korea. *Environmental Pollution, 270*, 116175.

Choy, C. A., & Drazen, J. C. (2013). Plastic for dinner? Observations of frequent debris ingestion by pelagic predatory fishes from the central North Pacific. *Marine Ecology – Progress Series, 485,* 155–163.

Cole, M., Lindeque, P., Fileman, E., Halsband, C., Goodhead, R., Moger, J., & Galloway, T. S. (2013). Microplastic ingestion by zooplankton. *Environmental Science and Technology, 47*(12), 6646–6655.

Cole, M., Lindeque, P., Fileman, E., Halsband, C., & Galloway, T. S. (2015). The impact of polystyrene microplastics on feeding, function and fecundity in the marine copepod Calanus helgolandicus. *Environmental Science and Technology, 49*(2), 1130–1137.

Compa, M., Ventero, A., Iglesias, M., & Deudero, S. (2018). Ingestion of microplastics and natural fibres in Sardina pilchardus (Walbaum, 1792) and Engraulis encrasicolus (Linnaeus, 1758) along the Spanish Mediterranean coast. *Marine Pollution Bulletin, 128,* 89–96.

Cózar, A., Echevarría, F., González-Gordillo, J. I., Irigoien, X., Úbeda, B., Hernández-León, S., ... & Duarte, C. M. (2014). Plastic debris in the open ocean. *Proceedings of the National Academy of Sciences, 111*(28), 10239–10244.

Daniel, D. B., Ashraf, P. M., & Thomas, S. N. (2020). Abundance, characteristics and seasonal variation of microplastics in Indian white shrimps (Fenneropenaeus indicus) from coastal waters off Cochin, Kerala, India. *Science of the Total Environment, 737,* 139839.

Davidson, K., & Dudas, S. E. (2016). Microplastic ingestion by wild and cultured Manila clams (Venerupisphilippinarum) from Baynes Sound, British Columbia. *Archives of Environmental Contamination and Toxicology, 71*(2), 147–156.

Davison, P., & Asch, R. G. (2011). Plastic ingestion by mesopelagic fishes in the North Pacific Subtropical gyre. *Marine Ecology – Progress Series, 432,* 173–180.

Dawson, A., Huston, W., Kawaguchi, S., King, C., Cropp, R., Wild, S., ... Bengtson Nash, S. (2018). Uptake and depuration kinetics influence microplastic bioaccumulation and toxicity in Antarctic krill (Euphausia superba). *Environmental Science and Technology, 52*(5), 3195–3201.

Détrée, C., & Gallardo-Escárate, C. (2018). Single and repetitive microplastics exposures induce immune system modulation and homeostasis alteration in the edible mussel Mytilus galloprovincialis. *Fish & Shellfish Immunology, 83,* 52–60.

Devriese, L. I., Van der Meulen, M. D., Maes, T., Bekaert, K., Paul-Pont, I., Frère, L., ... Vethaak, A. D. (2015). Microplastic contamination in brown shrimp (Crangoncrangon, Linnaeus 1758) from coastal waters of the southern North Sea and Channel area. *Marine Pollution Bulletin, 98*(1–2), 179–187.

Diaz-Basantes, M. F., Conesa, J. A., & Fullana, A. (2020). Microplastics in honey, beer, milk and refreshments in Ecuador as emerging contaminants. *Sustainability, 12*(14), 5514.

Digka, N., Tsangaris, C., Torre, M., Anastasopoulou, A., & Zeri, C. (2018). Microplastics in mussels and fish from the Northern Ionian Sea. *Marine Pollution Bulletin, 135,* 30–40.

Ding, J., Zhang, S., Razanajatovo, R. M., Zou, H., & Zhu, W. (2018). Accumulation, tissue distribution, and biochemical effects of polystyrene microplastics in the freshwater fish red tilapia (Oreochromis niloticus). *Environmental Pollution, 238,* 1–9.

Ding, H., Zhang, J., He, H., Zhu, Y., Dionysiou, D. D., Liu, Z., & Zhao, C. (2021). Do membrane filtration systems in drinking water treatment plants release Nano/microplastics? *Science of the Total Environment*, *755*(2), 142658.

Dris, R., Gasperi, J., Rocher, V., Saad, M., Renault, N., & Tassin, B. (2015). Microplastic contamination in an urban area: A case study in Greater Paris. *Environmental Chemistry*, *12*(5), 592–599.

Dowarah, K., Patchaiyappan, A., Thirunavukkarasu, C., Jayakumar, S., & Devipriya, S. P. (2020). Quantification of microplastics using Nile Red in two bivalve species Perna viridis and Meretrix meretrix from three estuaries in Pondicherry, India and microplastic uptake by local communities through bivalve diet. *Marine Pollution Bulletin*, *153*, 110982.

Eriksen, M., Lebreton, L. C., Carson, H. S., Thiel, M., Moore, C. J., Borerro, J. C., … Reisser, J. (2014). Plastic pollution in the world's oceans: More than 5 trillion plastic pieces weighing over 250,000 tons afloat at sea. *PLOS ONE*, *9*(12), e111913.

Fadare, O. O., Okoffo, E. D., & Olasehinde, E. F. (2021). Microparticles and microplastics contamination in African table salts. *Marine Pollution Bulletin*, *164*, 112006.

Fang, C., Zheng, R., Chen, H., Hong, F., Lin, L., Lin, H., … Bo, J. (2019). Comparison of microplastic contamination in fish and bivalves from two major cities in Fujian Province, China and the implications for human health. *Aquaculture*, *512*, 734322.

Feng, D., Yuan, H., Tang, J., Cai, X., & Yang, B. (2021). Preliminary investigation of microplastics in the production process of sea salt sourced from the Bohai Sea, China, using an optimised and consistent approach. *Food Additives & Contaminants: Part A*, *38*(12), 2151–2164.

Feng, Z., Zhang, T., Li, Y., He, X., Wang, R., Xu, J., & Gao, G. (2019). The accumulation of microplastics in fish from an important fish farm and mariculture area, Haizhou Bay, China. *Science of the Total Environment*, *696*, 133948.

Ferreira, P., Fonte, E., Soares, M. E., Carvalho, F., & Guilhermino, L. (2016). Effects of multi-stressors on juveniles of the marine fish Pomatoschistusmicrops: gold nanoparticles, microplastics and temperature. *Aquatic Toxicology*, *170*, 89–103.

Forbes, G., Gleich, C., & Brotemarkle, K. (2019). Microplastics in fiddler crabs. https://digitalcommons.coastal.edu/cgi/viewcontent.cgi?article=1340&context=honors-theses.

Fossi, M. C., Coppola, D., Baini, M., Giannetti, M., Guerranti, C., Marsili, L., … Clò, S. (2014). Large filter feeding marine organisms as indicators of microplastic in the pelagic environment: The case studies of the Mediterranean basking shark (Cetorhinus maximus) and fin whale (Balaenoptera physalus). *Marine Environmental Research*, *100*, 17–24.

Gardon, T., Reisser, C., Soyez, C., Quillien, V., & Le Moullac, G. (2018). Microplastics affect energy balance and gametogenesis in the pearl oyster Pinctada margaritifera. *Environmental Science & Technology*, *52*(9), 5277–5286.

Garnier, Y., Jacob, H., Guerra, A. S., Bertucci, F., & Lecchini, D. (2019). Evaluation of microplastic ingestion by tropical fish from Moorea Island, French Polynesia. *Marine Pollution Bulletin*, *140*, 165–170.

Gaspar, T. R., Chi, R. J., Parrow, M. W., & Ringwood, A. H. (2018). Cellular bio-reactivity of micro-and nano-plastic particles in oysters. *Frontiers in Marine Science*, *5*, 345.

Ghosh, G. C., Akter, S. M., Islam, R. M., Habib, A., Chakraborty, T. K., Zaman, S., … & Wahid, M. A. (2021). Microplastics contamination in commercial marine fish from the Bay of Bengal. *Regional Studies in Marine Science*, *44*, 101728.

GoI (2017) *Annual Report 2016–17* (pp. 1–109). Salt Department, Ministry of Commerce & Industry, Government of India.

Gomiero, A., Strafella, P., Øysæd, K. B., & Fabi, G. (2019). First occurrence and com-position assessment of microplastics in native mussels collected from coastal and offshore areas of the northern and central Adriatic Sea. *Environmental Science and Pollution Research*, *26*(24), 24407–24416.

Goswami, P., Vinithkumar, N. V., & Dharani, G. (2020). First evidence of micro-plastics bioaccumulation by marine organisms in the Port Blair Bay, Andaman Islands. *Marine Pollution Bulletin*, *155*, 111163.

Graham, P., Palazzo, L., de Lucia, G. A., Telfer, T. C., Baroli, M., & Carboni, S. (2019). Microplastics uptake and egestion dynamics in Pacific oysters, Magallana gigas (Thunberg, 1793), under controlled conditions. *Environmental Pollution*, *252*(A), 742–748.

Green, D. S., Boots, B., Sigwart, J., Jiang, S., & Rocha, C. (2016). Effects of con-ventional and biodegradable microplastics on a marine ecosystem engineer (Arenicola marina) and sediment nutrient cycling. *Environmental Pollution*, *208*, 426–434.

Green, D. S. (2016). Effects of microplastics on European flat oysters, Ostrea edulis and their associated benthic communities. *Environmental Pollution*, *216*, 95–103.

Green, D. S., Boots, B., O'Connor, N. E., & Thompson, R. (2017). Microplastics affect the ecological functioning of an important biogenic habitat. *Environmental Science & Technology*, *51*(1), 68–77.

Green, D. S., Kregting, L., Boots, B., Blockley, D. J., Brickle, P., Da Costa, M., & Crowley, Q. (2018). A comparison of sampling methods for seawater microplas-tics and a first report of the microplastic litter in coastal waters of Ascension and Falkland Islands. *Marine Pollution Bulletin*, *137*, 695–701.

Green, D. S., Colgan, T. J., Thompson, R. C., & Carolan, J. C. (2019). Exposure to microplastics reduces attachment strength and alters the haemolymph proteome of blue mussels (Mytilus edulis). *Environmental Pollution*, *246*, 423–434.

Guilhermino, L., Vieira, L. R., Ribeiro, D., Tavares, A. S., Cardoso, V., Alves, A., & Almeida, J. M. (2018). Uptake and effects of the antimicrobial florfenicol, microplastics and their mixtures on freshwater exotic invasive bivalve Corbicula fluminea. *Science of the Total Environment*, *622*, 1131–1142.

Gündoğdu, S. (2018). Contamination of table salts from Turkey with microplastics. *Food Additives & Contaminants: Part A*, *35*(5), 1006–1014.

Halstead, J. E., Smith, J. A., Carter, E. A., Lay, P. A., & Johnston, E. L. (2018). Assessment tools for microplastics and natural fibres ingested by fish in an urbanised estuary. *Environmental Pollution*, *234*, 552–561.

Harit A. K., Aiswyira V. P., Babu, N., Sruthi S Nair, & Ramasamy E. V. (2019). Microplastic contamination in green mussels: a favourite seafood in Kerala, India. In AdCoRe 2019, National Centre for Coastal Research, Chennai. December 17–19, 2019.

Horn, D., Miller, M., Anderson, S., & Steele, C. (2019). Microplastics are ubiquitous on California beaches and enter the coastal food web through consumption by Pacific mole crabs. *Marine Pollution Bulletin, 139*, 231–237.

Horton, A. A., Jürgens, M. D., Lahive, E., van Bodegom, P. M., & Vijver, M. G. (2018). The influence of exposure and physiology on microplastic ingestion by the freshwater fish Rutilus rutilus (roach) in the River Thames, UK. *Environmental Pollution, 236*, 188–194.

Hossain, M. S., Sobhan, F., Uddin, M. N., Sharifuzzaman, S. M., Chowdhury, S. R., Sarker, S., & Chowdhury, M. S. N. (2019). Microplastics in fishes from the Northern Bay of Bengal. *Science of the Total Environment, 690*, 821–830.

Hossain, M. S., Rahman, M. S., Uddin, M. N., Sharifuzzaman, S. M., Chowdhury, S. R., Sarker, S., & Chowdhury, M. S. N. (2020). Microplastic contamination in Penaeid shrimp from the northern Bay of Bengal. *Chemosphere, 238*, 124688.

Iñiguez, M. E., Conesa, J. A., & Fullana, A. (2017). Microplastics in Spanish table salt. *Scientific Reports, 7*(1), 1–7.

Jabeen, K., Su, L., Li, J., Yang, D., Tong, C., Mu, J., & Shi, H. (2017). Microplastics and mesoplastics in fish from coastal and fresh waters of China. *Environmental Pollution, 221*, 141–149.

James, K., Vasant, K., Padua, S., Gopinath, V., Abilash, K. S., Jeyabaskaran, R., ... & John, S. (2020). An assessment of microplastics in the ecosystem and selected commercially important fishes off Kochi, south eastern Arabian Sea, India. *Marine Pollution Bulletin, 154*, 111027.

Jonathan, M. P., Sujitha, S. B., Rodriguez-Gonzalez, F., Villegas, L. E. C., Hernández-Camacho, C. J., & Sarkar, S. K. (2021). Evidences of microplastics in diverse fish species off the Western Coast of Pacific Ocean, Mexico. *Ocean & Coastal Management, 204*, 105544.

Karami, A., Romano, N., Galloway, T., & Hamzah, H. (2016). Virgin microplastics cause toxicity and modulate the impacts of phenanthrene on biomarker responses in African catfish (Clariasgariepinus). *Environmental Research, 151*, 58–70.

Karami, A., Golieskardi, A., Choo, C. K., Romano, N., Ho, Y. B., & Salamatinia, B. (2017). A high-performance protocol for extraction of microplastics in fish. *Science of the Total Environment, 578*, 485–494.

Karami, A., Golieskardi, A., Choo, C. K., Larat, V., Galloway, T. S., & Salamatinia, B. (2017a). The presence of microplastics in commercial salts from different countries. *Scientific Reports, 7*, 46173.

Karbalaei, S., Golieskardi, A., Hamzah, H. B., Abdulwahid, S., Hanachi, P., Walker, T. R., & Karami, A. (2019). Abundance and characteristics of microplastics in commercial marine fish from Malaysia. *Marine Pollution Bulletin, 148*, 5–15.

Karuppasamy, P. K., Ravi, A., Vasudevan, L., Elangovan, M. P., Mary, P. D., Vincent, S. G., & Palanisami, T. (2020). Baseline survey of micro and mesoplastics in the gastro-intestinal tract of commercial fish from Southeast coast of the Bay of Bengal. *Marine Pollution Bulletin, 153*, 110974.

Keshavarzifard, M., Vazirzadeh, A., & Sharifinia, M. (2021). Occurrence and characterization of microplastics in white shrimp, Metapenaeusaffinis, living in a habitat highly affected by anthropogenic pressures, northwest Persian Gulf. *Marine Pollution Bulletin, 169*, 112581.

Kim, J. S., Lee, H. J., Kim, S. K., & Kim, H. J. (2018). Global pattern of microplastics (MPs) in commercial food-grade salts: sea salt as an indicator of seawater MP pollution. *Environmental Science & Technology*, *52*(21), 12819–12828.

Kosuth, M., Mason, S. A., & Wattenberg, E. V. (2018). Anthropogenic contamination of tap water, beer, and sea salt. *PloS One*, *13*(4), e0194970.

Kumar, V. E., Ravikumar, G., & Jeyasanta, K. I. (2018). Occurrence of microplastics in fishes from two landing sites in Tuticorin, South east coast of India. *Marine Pollution Bulletin*, *135*, 889–894.

Lee, H., Kunz, A., Shim, W. J., & Walther, B. A. (2019). Microplastic contamination of table salts from Taiwan, including a global review. *Scientific Reports*, *9*(1), 1–9.

Lefebvre, C., Saraux, C., Heitz, O., Nowaczyk, A., & Bonnet, D. (2019). Microplastics FTIR characterisation and distribution in the water column and digestive tracts of small pelagic fish in the Gulf of Lions. *Marine Pollution Bulletin*, *142*, 510–519.

Li, B., Su, L., Zhang, H., Deng, H., Chen, Q., & Shi, H. (2020a). Microplastics in fishes and their living environments surrounding a plastic production area. *Science of the Total Environment*, 138662.

Li, J., Yang, D., Li, L., Jabeen, K., & Shi, H. (2015). Microplastics in commercial bivalves from China. *Environmental Pollution*, *207*, 190–195.

Li, J., Green, C., Reynolds, A., Shi, H., & Rotchell, J. M. (2018). Microplastics in mussels sampled from coastal waters and supermarkets in the United Kingdom. *Environmental Pollution*, *241*, 35–44.

Li, L., Liu, D., Li, Z., Song, K., & He, Y. (2020). Evaluation of microplastic polyvinyl-chloride and antibiotics tetracycline co-effect on the partial nitrification process. *Marine Pollution Bulletin*, *160*, 111671.

Li, Q., Sun, C., Wang, Y., Cai, H., Li, L., Li, J., & Shi, H. (2019). Fusion of microplastics into the mussel byssus. *Environmental Pollution*, *252*, 420–426.

Liebezeit, G., & Liebezeit, E. (2013). Non-pollen particulates in honey and sugar. *Food Additives & Contaminants: Part A*, *30*(12), 2136–2140.

Lusher, A. L., Mchugh, M., & Thompson, R. C. (2013). Occurrence of microplastics in the gastrointestinal tract of pelagic and demersal fish from the English Channel. *Marine Pollution Bulletin*, *67*(1–2), 94–99.

Lusher, A. L., Tirelli, V., O'Connor, I., & Officer, R. (2015). Microplastics in Arctic polar waters: The first reported values of particles in surface and sub-surface samples. *Scientific Reports*, *5*(1), 1–9.

Lusher, A. (2015). Microplastics in the marine environment: distribution, interactions and effects. In Bergmann, M., Gutow, L., & Klages, M. (Eds.) *Marine Anthropogenic Litter* (pp. 245–307). Springer Open.

Lv, W., Zhou, W., Lu, S., Huang, W., Yuan, Q., Tian, M., ... He, D. (2019). Microplastic pollution in rice-fish co-culture system: A report of three farmland stations in Shanghai, China. *Science of the Total Environment*, *652*, 1209–1218.

Maaghloud, H., Houssa, R., Ouansafi, S., Bellali, F., El Bouqdaoui, K., Charouki, N., & Fahde, A. (2020). Ingestion of microplastics by pelagic fish from the Moroccan Central Atlantic coast. *Environmental Pollution*, *261*, 114194.

Magni, S., Gagné, F., André, C., Della Torre, C., Auclair, J., Hanana, H., ... & Binelli, A. (2018). Evaluation of uptake and chronic toxicity of virgin polystyrene micro-beads in freshwater zebra mussel Dreissena polymorpha (Mollusca: Bivalvia). *Science of the Total Environment*, *631*, 778–788.

Mishra, S., charan Rath, C., & Das, A. P. (2019). Marine microfiber pollution: A review on present status and future challenges. *Marine Pollution Bulletin*, *140*, 188–197.

Morgana, S., Ghigliotti, L., Estévez-Calvar, N., Stifanese, R., Wieckzorek, A., Doyle, T., ... & Garaventa, F. (2018). Microplastics in the Arctic: A case study with sub-surface water and fish samples off Northeast Greenland. *Environmental Pollution*, *242*, 1078–1086.

Mühlschlegel, P., Hauk, A., Walter, U., & Sieber, R. (2017). Lack of evidence for microplastic contamination in honey. *Food Additives & Contaminants: Part A*, *34*(11), 1982–1989.

Naidu, S. A. (2019). Preliminary study and first evidence of presence of microplastics and colorants in green mussel, Perna viridis (Linnaeus, 1758), from southeast coast of India. *Marine Pollution Bulletin*, *140*, 416–422.

Naidu, S. A., Ranga Rao, V., & Ramu, K. (2018). Microplastics in the benthic invertebrates from the coastal waters of Kochi, southeastern Arabian Sea. *Environmental Geochemistry and Health*, *40*(4), 1377–1383.

Naji, A., Nuri, M., & Vethaak, A. D. (2018). Microplastics contamination in molluscs from the northern part of the Persian Gulf. *Environmental Pollution*, *235*, 113–120.

Neves, D., Sobral, P., Ferreira, J. L., & Pereira, T. (2015). Ingestion of microplastics by commercial fish off the Portuguese coast. *Marine Pollution Bulletin*, *101*(1), 119–126.

Nie, H., Wang, J., Xu, K., Huang, Y., & Yan, M. (2019). Microplastic pollution in water and fish samples around Nanxun Reef in Nansha Islands, South China Sea. *Science of the Total Environment*, *696*, 134022.

Nithin, A., Sundaramanickam, A., Surya, P., Sathish, M., Soundharapandiyan, B., & Balachandar, K., 2021. Microplastic contamination in salt pans and commercial salts: A baseline study on the salt pans of Marakkanam and Parangipettai, Tamil Nadu, India. *Marine Pollution Bulletin*, *165*, 112101.

Obbard, R. W., Sadri, S., Wong, Y. Q., Khitun, A. A., Baker, I., & Thompson, R. C. (2014). Global warming releases microplastic legacy frozen in Arctic Sea ice. *Earth's Future*, *2*(6), 315–320.

Ory, N., Chagnon, C., Felix, F., Fernández, C., Ferreira, J. L., Gallardo, C., ... & Mojica, H. (2018). Low prevalence of microplastic contamination in planktivorous fish species from the southeast Pacific Ocean. *Marine Pollution Bulletin*, *127*, 211–216.

Paul-Pont, I., Lacroix, C., Fernández, C. G., Hégaret, H., Lambert, C., Le Goïc, N., ... & Guyomarch, J. (2016). Exposure of marine mussels Mytilus spp. to polystyrene microplastics: toxicity and influence on fluoranthene bioaccumulation. *Environmental Pollution*, *216*, 724–737.

Pittura, L., Avio, C. G., Giuliani, M. E., d'Errico, G., Keiter, S. H., Cormier, B., ... & Regoli, F. (2018). Microplastics as vehicles of environmental PAHs to marine organisms: combined chemical and physical hazards to the Mediterranean mussels, Mytilus galloprovincialis. *Frontiers in Marine Science*, *5*, 103.

Pozo, K., Gomez, V., Torres, M., Vera, L., Nuñez, D., Oyarzún, P., ... & Přibylová, P. (2019). Presence and characterization of microplastics in fish of commercial importance from the Biobío region in central Chile. *Marine Pollution Bulletin*, *140*, 315–319.

Qiao, R., Deng, Y., Zhang, S., Wolosker, M. B., Zhu, Q., Ren, H., & Zhang, Y. (2019). Accumulation of different shapes of microplastics initiates intestinal injury and gut microbiota dysbiosis in the gut of zebrafish. *Chemosphere*, *236*, 124334.

Qu, X., Su, L., Li, H., Liang, M., & Shi, H. (2018). Assessing the relationship between the abundance and properties of microplastics in water and in mussels. *Science of the Total Environment*, *621*, 679–686.

Ramasamy, E. V., Sruthi S N, Harit, A. K., & Babu, N., (2019). Microplastics in human consumption: table salt contaminated with microplastics. In S. Babel, A. Haarstrick, M. S. Babel, & A. Sharp (Eds.), *Microplastics in the Water Environment* (pp. 74–80). Cuvillier Verlag.

Ramasamy, E. V., Sruthy, S., Harit, A. K., Mohan, M., & Binish, M. B. (2021). Microplastic pollution in the surface sediment of Kongsfjorden, Svalbard, Arctic. *Marine Pollution Bulletin*, *173*, 112986.

Renzi, M., & Blašković, A. (2018). Litter & microplastics features in table salts from marine origin: Italian versus Croatian brands. *Marine Pollution Bulletin*, *135*, 62–68.

Revel, M., Lagarde, F., Perrein-Ettajani, H., Bruneau, M., Akcha, F., Sussarellu, R., ... & Châtel, A. (2019). Tissue-specific biomarker responses in the blue mussel Mytilus spp. exposed to a mixture of microplastics at environmentally relevant concentrations. *Frontiers in Environmental Science*, *7*, 33.

Ribeiro, F., Garcia, A. R., Pereira, B. P., Fonseca, M., Mestre, N. C., Fonseca, T. G., ... & Bebianno, M. J. (2017). Microplastics effects in Scrobicularia plana. *Marine Pollution Bulletin*, *122*(1–2), 379–391.

Rist, S. E., Assidqi, K., Zamani, N. P., Appel, D., Perschke, M., Huhn, M., & Lenz, M. (2016). Suspended micro-sized PVC particles impair the performance and decrease survival in the Asian green mussel Perna viridis. *Marine Pollution Bulletin*, *111*(1–2), 213–220.

Rochman, C. M., Tahir, A., Williams, S. L., Baxa, D. V., Lam, R., Miller, J. T., ... Teh, S. J. (2015). Anthropogenic debris in seafood: Plastic debris and fibers from textiles in fish and bivalves sold for human consumption. *Scientific Reports*, *5*(1), 1–10.

Santana, M. F., Moreira, F. T., Pereira, C. D., Abessa, D. M., & Turra, A. (2018). Continuous exposure to microplastics does not cause physiological effects in the cultivated mussel Perna perna. *Archives of Environmental Contamination and Toxicology*, *74*(4), 594–604.

Sathish, M. N., Jeyasanta, I., & Patterson, J. (2020). *Microplastics in salt of Tuticorin, southeast coast of India. Archives of Environmental Contamination and Toxicology*, *79*(1), 111–121.

Santana, M. F., Moreira, F. T., Pereira, C. D., Abessa, D., & Turra, A. (2018). Continuous exposure to microplastics does not cause physiological effects in the cultivated mussel Pernaperna. *Archives of Environmental Contamination and Toxicology*, *74*(4), 594–604.

Santos, J. D., & Jobling, M. (1992). A model to describe gastric evacuation in cod (Gadus morhua L.) fed natural prey. *ICES Journal of Marine Science*, *49*(2), 145–154.

Scott, N., Porter, A., Santillo, D., Simpson, H., Lloyd-Williams, S., & Lewis, C. (2019). Particle characteristics of microplastics contaminating the mussel Mytilus edulis and their surrounding environments. *Marine Pollution Bulletin*, *146*, 125–133.

Selvam, S., Manisha, A., Venkatramanan, S., Chung, S. Y., Paramasivam, C. R., & Singaraja, C. (2020). Microplastic presence in commercial marine sea salts: A baseline study along Tuticorin Coastal salt pan stations, Gulf of Mannar, South India. *Marine Pollution Bulletin, 150*, 110675.

Seth, C. K., & Shriwastav, A. (2018). Contamination of Indian sea salts with microplastics and a potential prevention strategy. *Environmental Science and Pollution Research, 25*(30), 30122–30131.

Silva-Cavalcanti, J. S., Silva, J. D. B., de França, E. J., de Araújo, M. C. B., & Gusmao, F. (2017). Microplastics ingestion by a common tropical freshwater fishing resource. *Environmental Pollution, 221*, 218–226.

Sivagami, M., Selvambigai, M., Devan, U., Velangani, A. A. J., Karmegam, N., Biruntha, M., Arun, A., Kim, W., Govarthanan, M., & Kumar, P. (2021). Extraction of microplastics from commonly used sea salts in India and their toxicological evaluation. *Chemosphere, 263*, 128181.

Slootmaekers, B., Carteny, C. C., Belpaire, C., Saverwyns, S., Fremout, W., Blust, R., & Bervoets, L. (2019). Microplastic contamination in gudgeons (Gobio gobio) from Flemish rivers (Belgium). *Environmental Pollution, 244*, 675–684.

Su, L., Nan, B., Hassell, K. L., Craig, N. J., & Pettigrove, V. (2019). Microplastics biomonitoring in Australian urban wetlands using a common noxious fish (Gambusia holbrooki). *Chemosphere, 228*, 65–74.

Su, L., Deng, H., Li, B., Chen, Q., Pettigrove, V., Wu, C., & Shi, H. (2019a). The occurrence of microplastic in specific organs in commercially caught fishes from coast and estuary area of east China. *Journal of Hazardous Materials, 365*, 716–724.

Sun, D., Wang, J., Xie, S., Tang, H., Zhang, C., Xu, G., ... & Zhou, A. (2021). Characterization and spatial distribution of microplastics in two wild captured economic freshwater fish from north and west rivers of Guangdong province. *Ecotoxicology and Environmental Safety, 207*, 111555.

Sussarellu, R., Suquet, M., Thomas, Y., Lambert, C., Fabioux, C., Pernet, M. E. J., ... & Corporeau, C. (2016). Oyster reproduction is affected by exposure to polystyrene microplastics. *Proceedings of the National Academy of Sciences, 113*(9), 2430–2435.

Tanaka, K., & Takada, H. (2016). Microplastic fragments and microbeads in digestive tracts of planktivorous fish from urban coastal waters. *Scientific Reports, 6*, 34351.

Teng, J., Wang, Q., Ran, W., Wu, D., Liu, Y., Sun, S., ... & Zhao, J. (2019). Microplastic in cultured oysters from different coastal areas of China. *Science of the Total Environment, 653*, 1282–1292.

Thiagarajan, V., Alex, S. A., Seenivasan, R., Chandrasekaran, N., & Mukherjee, A. (2021). Toxicity evaluation of nano-TiO2 in the presence of functionalized microplastics at two trophic levels: Algae and crustaceans. *Science of the Total Environment, 784*, 147262.

Van Cauwenberghe, L., Claessens, M., Vandegehuchte, M. B., & Janssen, C. R. (2015). Microplastics are taken up by mussels (Mytilus edulis) and lugworms (Arenicola marina) living in natural habitats. *Environmental Pollution, 199*, 10–17.

Van Cauwenberghe, L., & Janssen, C. R. (2014). Microplastics in bivalves cultured for human consumption. *Environmental Pollution, 193*, 65–70.

Van Cauwenberghe, L., Vanreusel, A., Mees, J., & Janssen, C. R. (2013). Microplastic pollution in deep-sea sediments. *Environmental Pollution, 182*, 495–499.

Vidyasakar, A., Krishnakumar, S., Kumar, K. S., Neelavannan, K., Anbalagan, S., Kasilingam, K., Srinivasalu, S., Saravanan, P., Kamaraj, S., & Magesh, N. S. (2021). Microplastic contamination in edible sea salt from the largest salt-producing states of India. *Marine Pollution Bulletin, 171*, 112728.

Von Moos, N., Burkhardt-Holm, P., & Köhler, A. (2012). Uptake and effects of microplastics on cells and tissue of the blue mussel Mytilus edulis L. after an experimental exposure. *Environmental Science & Technology, 46*(20), 11327–11335.

Wang, Z. M., Wagner, J., Ghosal, S., Bedi, G., & Wall, S. (2017). SEM/EDS and optical microscopy analyses of microplastics in ocean trawl and fish guts. *Science of the Total Environment, 603*, 616–626.

Wagner, J., Wang, Z. M., Ghosal, S., Rochman, C., Gassel, M., & Wall, S. (2017). Novel method for the extraction and identification of microplastics in ocean trawl and fish gut matrices. *Analytical Methods, 9*(9), 1479–1490.

Watts, A. J., Lewis, C., Goodhead, R. M., Beckett, S. J., Moger, J., Tyler, C. R., & Galloway, T. S. (2014). Uptake and retention of microplastics by the shore crab Carcinusmaenas. *Environmental Science and Technology, 48*(15), 8823–8830.

Webb, S., Ruffell, H., Marsden, I., Pantos, O., & Gaw, S. (2019). Microplastics in the New Zealand green lipped mussel Perna canaliculus. *Marine Pollution Bulletin, 149*, 110641.

Wegner, A., Besseling, E., Foekema, E. M., Kamermans, P., & Koelmans, A. A. (2012). Effects of nanopolystyrene on the feeding behavior of the blue mussel (Mytilus edulis L.). *Environmental Toxicology and Chemistry, 31*(11), 2490–2497.

Wieczorek, A. M., Morrison, L., Croot, P. L., Allcock, A. L., MacLoughlin, E., Savard, O., ... & Doyle, T. K. (2018). Frequency of microplastics in mesopelagic fishes from the Northwest Atlantic. *Frontiers in Marine Science, 5*, 39.

Xu, X. Y., Lee, W. T., Chan, A. K. Y., Lo, H. S., Shin, P. K. S., & Cheung, S. G. (2017). Microplastic ingestion reduces energy intake in the clam Atactodea striata. *Marine Pollution Bulletin, 124*(2), 798–802.

Yang, D., Shi, H., Li, L., Li, J., Jabeen, K., & Kolandhasamy, P. (2015). Microplastic pollution in table salts from China. *Environmental Science & Technology, 49*(22), 13622–13627.

Yaranal, N. A., Subbiah, S., & Mohanty, K. (2021). Identification, extraction of microplastics from edible salts and its removal from contaminated seawater. *Environmental Technology & Innovation, 21*, 101253.

Yin, L., Jiang, C., Wen, X., Du, C., Zhong, W., Feng, Z., ... Ma, Y. (2019). Microplastic pollution in surface water of urban lakes in Changsha, China. *International Journal of Environmental Research and Public Health, 16*(9), 1650.

Yuan, W., Liu, X., Wang, W., Di, M., & Wang, J. (2019). Microplastic abundance, distribution and composition in water, sediments, and wild fish from Poyang Lake, China. *Ecotoxicology and Environmental Safety, 170*, 180–187.

Zhang, F., Man, Y. B., Mo, W. Y., Man, K. Y., & Wong, M. H. (2020). Direct and indirect effects of microplastics on bivalves, with a focus on edible species: A mini-review. *Critical Reviews in Environmental Science and Technology, 50*(20), 2109–2143.

Zhang, J., Tian, K., Lei, C., & Min, S. (2018). Identification and quantification of microplastics in table sea salts using micro-NIR imaging methods. *Analytical Methods, 10*(24), 2881–2887.

Zhang, R., Zhang, Y., Antila, H. S., Lutkenhaus, J. L., & Sammalkorpi, M. (2017). Role of salt and water in the plasticization of PDAC/PSS polyelectrolyte assemblies. *The Journal of Physical Chemistry B*, *121*(1), 322–333.

Zhang, S., Ding, J., Razanajatovo, R. M., Jiang, H., Zou, H., & Zhu, W. (2019). Interactive effects of polystyrene microplastics and roxithromycin on bioaccumulation and biochemical status in the freshwater fish red tilapia (Oreochromis niloticus). *Science of the total environment*, *648*, 1431–1439.

Zhang, L., Xie, Y., Zhong, S., Liu, J., Qin, Y., & Gao, P. (2021). Microplastics in freshwater and wild fishes from Lijiang River in Guangxi, Southwest China. *Science of the Total Environment*, *755*, 142428.

Zhu, L., Wang, H., Chen, B., Sun, X., Qu, K., & Xia, B. (2019). Microplastic ingestion in deep-sea fish from the South China Sea. *Science of the Total Environment*, *677*, 493–501.

Microplastics in Potable Water and Beverages

3

3.1 INTRODUCTION

Microplastic contamination has been extensively reported in natural water bodies from the tropics to polar waters. However, studies on the presence of microplastic (MP) particles in potable water such as tap water supplied through public distribution systems, bottled water including mineral water, well water (groundwater used for all purposes) and other beverages being consumed by humans are under-reported when compared with natural water bodies. Despite the fact that drinking water occupies a major portion of the human diet by supplying required minerals and trace nutrients, the reports on MP contamination of potable water remain limited. The research publications on MP contamination of drinking water are summarized in Table 3.1. Most of the studies are reported from Germany; only one study is from India (Ganesan et al., 2019).

Apart from drinking water, beer, the predominant beverage being consumed, has also been reported to have MP contamination. Like potable water, the number of studies reporting MPs in beer is very few and most of the studies are from Germany, followed by Mexico and the USA (Table 3.2). Other beverages reported with MP contamination include white wines, cold tea, soft drinks, energy drinks and skimmed milk (Shruti et al., 2020a; Prata et al., 2020; Diaz-Basantes et al., 2020). This chapter will focus on the studies reporting MP contamination in potable water, beer and other beverages for human consumption and the significance of such contamination in detail.

DOI: 10.1201/9781003201755-3

TABLE 3.1   Microplastics in Human Consumable Water: Bottled Water, Drinking Water and Tap Water

SAMPLE TYPE	COUNTRY	SHAPE	MICROPLASTIC PARTICLE/L	COLOR	POLYMER TYPE	REFERENCES
				MICROPLASTICS		
Bottled water	China	Fibers, fragments	2–23	–	PP, PS, PE, PET, PU, PVC, PAA, PAAM, PEVA, CE	Zhou et al., 2021
	Thailand	Fibers, fragments	140 (SUP*) Glass bottle 52 (GB*)	Transparent, blue, brown and reddish brown	PE, PET, PP, PA, PVC, PVF, PTFE, PMMA	Kankanige and Babel, 2021
	UK	–	0.00011	–	PS, ABS	Johnson et al., 2020
	Italy	–	5.42	–	–	Zuccarello et al., 2019
	Germany	Fragments	2–44 (SUP) 28–241 (RPB*) 4–156 (GB)	–	PEST, PE, PA, PE, PP, paraffin	Schymanski et al., 2018
	samples from Nine countries	Fragments, films, fibers, foams, pellets	average 325	–	PP, nylon, PS, PE, PEST,	Mason et al., 2018

(Continued)

TABLE 3.1 (CONTINUED) Microplastics in Human Consumable Water: Bottled Water, Drinking Water and Tap Water

| SAMPLE TYPE | COUNTRY | SHAPE | MICROPLASTICS | | | |
			MICROPLASTIC PARTICLE/L	COLOR	POLYMER TYPE	REFERENCES
	Germany	–	*Single-use PET* 2,649 (SUP) *Reusable PET* 4,889 (RPB) *Glass bottle* 6,292 (GB)	–	*Single-use and reusable PET* PP, PE, PET + olefin, PET *Glass bottle* PP, PE, PET, styrene-butadiene-copolymer	Oßmann et al., 2018
	Germany	Fibers	30–57	–	Cellulose, PE, PS	Wiesheu et al., 2016
Drinking water Groundwater/ treated water)	China	Fibers, fragments, spheres	351.9 338–400 (Treated Water)	–	PP, PS, PET, PVC, PA, PMMA PE, PP, PET, PVC, PA	Shen et al., 2021
	Brazil	Fibers, pellets, films	105.8 (Treated water)	–	–	Ferraz et al., 2020
	Mexico	Fibers, fragments	5–91	Transparent, blue, red and green	PTT, epoxy resin	Shruti et al., 2020

(Continued)

TABLE 3.1 (CONTINUED) Microplastics in Human Consumable Water: Bottled Water, Drinking Water and Tap Water

| SAMPLE TYPE | COUNTRY | SHAPE | MICROPLASTICS |||| REFERENCES |
			MICROPLASTIC PARTICLE/L	COLOR	POLYMER TYPE	
	Germany	Fibers, fragments	*Groundwater* 0–7 MPs (Per m³) *Treated water* 0.7 MPs (Per m³)	–	PE, PA, PEST, PVC, epoxy resin	Mintenig et al., 2019
	India	Fibers, fragments	*Groundwater* 4–7 *Treated water* 2–6	White, blue, green, yellow, pink and black	PET, PA	Ganesan et al., 2019
	China	Fibers, fragments	0.2–0.7	Black, blue, gray and transparent	PET, PE, PVC, PAA, PDMS, PBT, PVS, PEVA, PPDS, PMPS, PVA, POM, PBO + MBI, PDAA, rayon	Zhang et al., 2019
	Czech Republic	Fragments, spheres, fibers	*Treated drinking water* 338–628	–	PAM, PE, PET, PP, PVC	Pivokonsky et al., 2018

(Continued)

TABLE 3.1 (CONTINUED) Microplastics in Human Consumable Water: Bottled Water, Drinking Water and Tap Water

SAMPLE TYPE	COUNTRY	SHAPE	MICROPLASTICS				
			MICROPLASTIC PARTICLE/L	COLOR	POLYMER TYPE	REFERENCES	
Tap water	China	Fibers, fragments, spheres	343.5	–	PE, PP, PET, PA, PVC	Shen et al., 2021	
	China	Fibers, fragments, spheres	0–1,247	–	PE, PP, PS, PE+PP, PET, PPS	Tong et al., 2020	
	China	Fibers, fragments	0.3–1.6	Black, blue, grey and transparent	PET, PE, PS, PMPS, PAA, PAM, PI, PDMS, rayon, polyester	Zhang et al., 2020	
	Germany	Fragments	0.0007	–	PA, PVC, PE, polyester, epoxy resin	Mintenig et al., 2019	
	Denmark	Fragments	0.58	Blue, pink and black	PET, PP, PS	Strand et al., 2018	
	Global	Fibers, fragments, films	0–61	Blue, red, pink and brown	–	Kosuth et al. 2018	

Notes: PPS = polyphenylene sulfite; PBT = polybutylene terephthalate; PAA = polyacrylic acid; PMPS = poly methyl phenyl siloxane; PVS = polyvinyl sulfide; PAM = polyacrylamide; PDMS = polydimethylsiloxane; PEVA = polyethylene and ethylene-vinyl acetate; PA = polyamide; PPDS = polyphenylene disulfide; POM = polyoxymethylene; PBO = polyphenylene benzobisoxazole; PVA = polyvinyl alcohol; PDAP = polydiallyl phthalate; PS = polystyrene; PI = polyisoprene; PCL = polycaprolactone; PET = polyethylene terephathalate; PMS = poly alpha-methylstyrene; PTFE = polytetrafluoroethylene; PC = polycarbonate; PMMA = poly methyl methacrylate; PBT = poly butylene terephthalate; PB = polybutene; PVC = polyvinylchloride; ABS = acrylonitrile butadiene styrene; CE = cellulose; PAAm = polyacrylamide; PEVA = polyethylene vinyl acetate; PTT = polytrimethylene terephthalate; PBT = polybutylacrylate; PEST = polyester + polyethylene terephthalate.

TABLE 3.2 Microplastics in Beer and White Wine

COUNTRY	SHAPE	MICROPLASTICS ABUNDANCE AND CHARACTERISTICS				
		PARTICLES/L (P/L)	SIZE (µm)	COLOR	POLYMER TYPE	REFERENCES
Studies on beer						
Germany	Fibers, fragments, granules	2–79 12–199 2–66	–	Blue, black, green and transparent	–	Liebezeit and Liebezeit, 2014
Germany	Fibers, fragments, granules	16 21 27	–	–	–	Lachenmeier et al., 2015
Germany	Fibers	–	1–5,000	–	PE, PS, cellulose	Wiesheu et al., 2016
USA	Fibers, fragments	0–14.3	100–5,000	Brown, black, blue, red and purple	–	Kosuth et al., 2018
Mexico	Fibers, fragments	0–28	100–3,000	Blue, black, brown and green	PA, PEA, PET, BP	Shruti et al., 2020a
Ecuador	Fibers, fragments	70–976	3.50–1,740.24	Red, blue and violet	PP, HDPE, LDPE, PAAm	Diaz-Basantes et al., 2020
Study on white wine						
Italy	–	79	–	–	PE	Prata et al., 2020

Notes:
PE = polyethylene; PET = polyethylene terephthalate; PA = polyamide; PEA = poly ester-amide; BP = blue pigment BP; PAAm = polyacrylamide; PP = polypropylene; HDPE = high-density polyethylene; LDPE = low-density polyethylene.

3.2 MICROPLASTICS IN POTABLE WATER, BEER AND WHITE WINE

As explained in the previous chapter, MPs were found in important human dietary items such as seafood, sea salt and honey. In addition, MPs are also witnessed in beverages including potable water, soft drinks, energy drinks and beer; however, the number of studies on this aspect is only a few.

3.2.1 Potable Water

Studies on MP contamination of potable water focus on different types/sources of water being consumed, such as well water, spring water, groundwater, public distribution water (centralized supply of municipal tap water by the government for drinking purposes) and commercially available bottled water.

Koelmans et al. (2019) have reviewed the global scenario of MP contamination in drinking water. The review indicates most detected polymers were PE, PET and PP while PMMA, PU, PS and PVC were found rarely. The review also reports on the different shapes of MPs found in global potable water, such as fragments, fibers, films, foams and pellets.

3.2.1.1 Microplastics in Groundwater

Panno et al. (2019) investigated the MP contamination in springs and well water of two Karst aquifers (Karst aquifers are a special type of fractured rock aquifers) in Illinois, USA, and found MPs as well as other anthropogenic contaminants in Karst groundwater systems. They have reported a maximum concentration of 15.2 particles/L and all the MPs were fibers in shape.

A study reported from China on the groundwater and drinking water supply chain determined a concentration of MPs ranging from 0 to 7 MPs m^{-3}. Polyethylene, polyamide, polyester and polyvinyl chloride were the polymers identified and the size of the MPs varied between 50 and 150 μm (Mintenig et al., 2019).

Pivokonsky et al. (2018) investigated the concentration of MPs in raw and treated drinking water in the Czech Republic. Raw and treated water samples were obtained from three drinking water treatment plants located in urban areas of the Czech Republic. The raw water of these treatment plants is supplied by diverse kinds of water bodies. Their study found MPs in all water samples; however, the abundance of MPs was higher in raw water (1,473 to 3,605 particles L^{-1}) than in treated water (338 to 628 particles L^{-1}). The majority of MPs were composed of polyethylene terephthalate (PET), polypropylene (PP) and poly-ethylene (PE) polymers and most of the particles were in the range of 1–10 µm. This is a rare study reporting microplastics down to the size of 1 µm; it is also mentioned in their findings that MPs smaller than 10 µm were predominant in both raw and treated water samples, accounting for up to 95%. The shapes of microparticles reported include fibers, spherical particles and fragments.

Strand et al. (2018) sampled drinking water from 17 sites around Denmark; 50 L of samples per site were collected directly from taps filtered through stainless steel filters (10 µm) into a closed steel filter system in order to prevent contamination. The findings of this study showed 30 MP-like particles per 50 L as the highest abundance; however, further analysis revealed only 3% of the MP-like particles were verified as MPs, while the rest of the particles were either cellulose-like material (76%), unknown (7%), protein-like (4%) or poor spectra (10%). The polymer profile of the MPs detected in the tap water samples included polyethylene terephthalate, polypropylene and polystyrene.

A study by Ganesan et al. (2019) analyzed surface water, groundwater and branded drinking water bottle samples collected from Chennai, Tamil Nadu, India, for the contamination of microplastics. The presence of MPs was highest in surface water followed by groundwater and bottled water. Most of the MPs were of fibers in shape, followed by fragments with varying colors including white, blue, green, yellow, pink and black as observed under an optical microscope. An SEM-EDX study on morphology and elemental analysis of microplastics indicated the presence of heavy metals such as Cr, Ti, Mo, Ba and Ru adhered to their surface. The polymers identified in the MPs were polyethylene terephthalate (PET) and polyamide (PA).

Shruti et al. (2020a) in their study from Mexico report the occurrence of microplastics in samples collected from free drinking water fountains located in 42 metro stations in Mexico City. MPs were observed in all the samples analyzed with an abundance ranging from 5 ± 2 to 91 ± 14 MPs L^{-1}, with an average of 18 ± 7 MPs L^{-1}. Fibers (transparent) were the predominant MPs (69%), followed by colored fibers: blue (24%) and red (7%). About 75% of total MPs were in the range of 0.1 to 1 mm in size. As per the micro-Raman spectroscopy analysis, MPs contained in the drinking water samples were mainly polyesters (poly-trimethylene terephthalate) and epoxy resin. The authors have suggested the possible contribution of wastewater discharges to the microplastic contamination of drinking water and indicated the free public drinking water fountains as potential microplastic hotspots for human consumption.

Ferraz et al. (2020) compared the concentration of MPs in the Sinos River water (source of drinking water supply) in southern Brazil prior to water treatment and the treated drinking water supplied to houses. The raw water was collected from the river at eight locations and the treated drinking water from eight residences. Water samples were processed with Nile Red and counted by fluorescence microscopy. Raw water from the river showed a higher load of MPs (an average of 330.2 particles L^{-1}) than treated water (105.8 particles L^{-1}). Fibers were the most abundant MP shape in the samples, suggesting that the effluent from washing machines could be a source of fibers along with untreated sewage contaminating the river.

3.2.1.2 Microplastics in Bottled Water

In Germany, 38 mineral waters (packed in 15 different returnable and 11 single-use plastic bottles, three beverage cartons and nine glass bottles) were analyzed for microplastic contamination. MP contamination was found in every type of water, between 5 and 20 μm in size (almost 80%). The average MP content was 118 ± 88 particles/l in returnables, 14 ± 14 particles/l in single-use plastic bottles, 11 ± 8 particles/l in beverage cartons and 0–253 particles/l in glass-bottled waters. The study also revealed the polymer of the MPs correlates with the materials the bottles/cartons were made-up of. Most of the particles in water from returnable plastic bottles were identified as polyester and polypropylene (Schymanski et al., 2018). However, Mason et al. (2018) investigated the MPs in 11 brands of bottled water from 19 locations in nine countries (Brazil, China, Germany, India, Indonesia, Lebanon, Mexico, Thailand, USA). Out of 259 samples examined, 242 bottled waters were contaminated with MPs. Fragments were the most common particles followed by fibers, while polypropylene was the most common polymer observed.

Both the studies (Schymanski et al., 2018; Mason et al., 2018) have observed that the source of MP contamination could be both the packaging and the bottling process itself or either of them. As the quantum of the consumption of bottled water across the globe is increasing, the results of the studies on MP contamination of bottled water warrant the need for further studies on the impacts of micro- and nanoplastics on human health.

Zhou et al. (2021) have studied 23 brands of bottled water in China for MP contamination. Their findings indicated the presence of MPs in the shapes of fibers and fragments; the size range was given as 0.025–5.0 mm and 2 to 23 particles/bottle were the range of abundance of the particles. They have also identified 11 polymer types in the MPs extracted from the bottled water, *viz.* polypropylene, polystyrene, polyethylene, polyethylene terephthalate, polyurethane, polyvinyl chloride, polyamide, polyacrylic acid, polyacrylamide, polyethylene vinyl acetate and cellulose. Based on their study, the researchers state that the source of MP contamination could be the raw water source and/or the packaging. This study also calculated the estimated daily intake (EDI) of microplastics by human beings by using the following formula:

$$EDI(MP/kg/d) = (C \times R/bw)$$

Where R stands for ingestion rate (L/d), C is equal to the concentration of MPs (particles/L) and bw mean body weight (kg). According to this calculation, the estimated daily intakes for adults and children were expected to be 0.274 MPs/kg/d and 0.600 MPs/kg/d (Zhou et al., 2021).

Another study on bottled water by Kankanige and Babel (2021) showed 140 ± 19 particles/L in single-use plastic-bottled water and 52 ± 4 particles/L in glass-bottled water. Fibers were predominant, forming 62.8% of the total particle count, followed by fragments. The polymer profile of the MPs extracted from bottled included PET, PE, PP and PA.

3.2.2 Beer and White Wine

Beer stands second to potable water in human consumption; however, only few studies have reported the occurrence of MPs in beer to date (Shruti et al., 2020b; Diaz-Basantes et al., 2020; Kosuth et al., 2018; Wiesheu et al., 2016; Lachenmeier et al., 2015; Liebezeit and Liebezeit, 2014). Only one study has reported MP contamination in white wine (Prata et al., 2020). Lachenmeier et al. (2015) and Liebezeit and Liebezeit (2014) investigated microplastic contamination in German beers while Kosuth et al. (2018) in American beers and Shruti et al. (2020a) in Mexican beers (Table 3.2). Kosuth et al. (2018) and Liebezeit and Liebezeit (2014) confirmed MP contamination in beers but Lachenmeier et al. (2015) did not agree with other researchers and reported that they were artefacts due to laboratory contamination. On the other hand, Liebezeit and Liebezeit (2014) have reported fibers, granules and fragments in beer at the following amounts: 2–79 fibers/L, 2–66 granules/L and 12–109 fragments/L, respectively. Meanwhile, 0–14.3 particles/L were reported by Kosuth et al. (2018) in American beer. The authors, highlighting the difference in beer processing between nations, have postulated that product processing is an important factor to understand anthropogenic contamination (Toussaint et al., 2019).

3.3 MICROPLASTICS IN SOFT DRINKS, COLD TEA AND ENERGY DRINKS

As explained in the previous chapter and in the preceding sections of the current chapter, the infiltration of microplastics (MPs) in human food items has attracted global concern due to the health risks involved with it. In this context, the reports on MP contamination of alcoholic beverages are few while

TABLE 3.3   Microplastics in Soft Drinks, Cold Tea and Energy Drinks from Mexico

		MICROPLASTICS ABUNDANCE AND CHARACTERISTICS			
SAMPLE TYPE	SHAPE	PARTICLES/L (P/L)	SIZE (μM)	COLOR	POLYMER TYPE
Cold tea	Fibers	1–6	100–2000	Blue, brown and red	PEA, BP, PA
Soft drinks	Fibers	0–7	100–3,000	Blue, brown and red	PEA, BP, PA, ABS
Energy drinks	Fibers	0–6	100–3,000	Blue and brown	PEA, BP, PA

Notes:
PEA = poly ester-amide;   BP = blue pigment;   ABS = acrylonitrile-butadiene-styrene; PA = polyamide.
Source: Shruti et al. (2020a).

non-alcoholic beverages like soft drinks, cold tea and energy drinks are even less studied. To date, only one study has reported MPs contamination in non-alcoholic beverages; in this study, Shruti et al. (2020a) analyzed samples from Mexico, and the beverage products examined include soft drinks (n = 19), energy drinks (n = 8), cold tea (n = 4) and the alcoholic beverage beer (n = 26). Their results indicated that 84% of the samples (48 out of 57 total samples) were contaminated with MPs. The abundance of MPs ranged from 11 to 40 in cold tea, soft drinks and energy drinks; however, MPs per liter in each category were quite less (Table 3.3). The trend of higher to lower abundance of MPs present among the samples (excluding beer) arrived by the authors is: soft drinks > energy drinks > cold tea. The MPs extracted from the non-alcoholic beverages were fibers in shape and the polymer content of the MPs was mostly polyamide (PA) and poly ester-amide (PEA).

3.3.1 Cold Tea

Microplastic particles were observed in all four samples studied. Even though the total average of MPs from cold tea samples was 11 ± 5.26, individual samples showed just 1 to 2 MPs/L except for one sample (out of the four) from which a maximum of 6 MPs/L could be identified. Only fiber-shaped MPs were noted in all samples and more than 80% of the MPs measured were < 1 mm in size (Shruti et al., 2020b).

3.3.2 Soft Drinks

Microplastic particles were found in all but three of the 19 samples tested. The total average of MPs identified was 40 ± 24.53. Only one sample showed a maximum of 7 MPs/L while other samples displayed a lesser abundance of MPs/L. All MPs identified from soft drink samples were fibers in shape with a size ranging from 0.1 to 3 mm, and particles measuring < 1 mm in size were dominant (Shruti et al., 2020a).

3.3.3 Energy Drinks

Five out of the eight samples analyzed showed the presence of MPs at a total average of 14 ± 5.79. Except for one sample, which exhibited 6 MPs/L, the other four samples were identified with just 2 MPs/L. Here also only fibers were detected and more than 70% of MPs identified measured < 1 mm in size (Shruti et al., 2020a).

3.4 POTENTIAL SOURCES OF MICROPLASTICS IN BEVERAGES

The beverage industry is a water-intensive industry and uses plenty of water in the production of beverages and in the process of cleaning the equipment, the packaging containers, both reusable and non-reusable (Shruti et al. 2021). This industry spends 3–4 L of freshwater for the production of 1 L of soft drinks (Grumezescu and Holban, 2018). The beverage or bottling industries use water from different sources in their production, cleaning and bottling processes; for instance, Kosuth et al. (2018) have reported that water from municipal water distribution systems is being used for beverage production in the USA. Other sources of water being used in this industry include surface water, public water networks, groundwater or rainwater; as described in the preceding sections of this chapter, many rivers, streams, groundwater and potable tap water/ fountains have been reported to have MPs contamination (Shruti et al., 2021; Johnson et al., 2020; Ganesan et al., 2019; Uhl et al., 2018). Thus, Sruthi et al. (2021) suggest that the water from different water bodies could be an important source of MP contamination in beverage products.

Packing materials, especially non-reusable or single-use containers, the lids used in such containers and improperly cleaned containers can also contribute to MP contamination in beverages, besides water (Schymanski et al., 2018; Sobhani et al., 2020).

The unpublished work of Aiswriya (2019) studied MP contamination in soft drinks and carbonated soda found more MP particles in carbonated soda, that is, an average of 120 ± 93.230 particles/L, while in soft drinks, 96 ± 40.515 particles/L were reported. In both cases, most of the particles are fragments in shape, followed by fibers. Polyethylene (PE), polyethylene terephthalate (PET), polypropylene (PP) and polyesters (PES) were the predominantly found polymers among the MP particles. George (2019) confirmed the presence of MPs in different types of drinking water samples (tap water, well water and commercially available bottled water). This study reported fragment shapes of MP particles as most abundant in tap water and bottled water, while fibers were abundant in well water. Polyvinyl chloride (PVC), polychloroprene (PCP), polystyrene (PS) and polyvinyl acetate (PVA) were the most detected types of polymers found in MPs of the bottled water; polypropylene (PP), polyvinyl acetate (PVA) and polystyrene (PS) were in the well water. Only polybutadiene (PB) was identified in the tap water. So, the studies of Aiswriya (2019) (unpublished data) confirm that the prevalence of MPs is more with the local brands (India-based) of soda, soft drinks and bottled water than the international brands, supplementing the views reported in the literature such as that water being used in the production process and in the cleaning of containers used for packaging could be potential sources of MP contamination in soft drinks and bottled water.

3.5 CONCLUSION

The beverage industry – both alcoholic and non-alcoholic – has a key role in the human diet; in a few nations, the per capita consumption of beverages exceeds the quantity of drinking water consumed daily (Shruti et al., 2020a). Thus, the contamination of MPs in beverages attracts global concern. Similar is the case with the potable water supply sector, including bottled water, which also plays a significant role in the human diet as well as health; hence the MP contamination of drinking water is also viewed as a serious issue. As far as the drinking water supply is concerned, the presence of a higher abundance of MPs was noted in the raw water prior to treatment; after treatment, a reduction in the number of MPs has been observed consistently by many authors. Mixing of treated or untreated sewage or wastewater apart from surface run-off has been the main source of MPs in the water source being used for drinking water supply. Most of the potable water from municipal supplies or fountains as well as bottled water has been contaminated predominantly with fibrous-shaped MPs. The presence of fibrous MPs in potable water has been attributed to the contamination of the source water with untreated laundry effluent. Similar to potable/bottled water, most of the beverages are also reported with predominantly fiber-shaped MPs. The polymer profile of the MPs extracted from beverages and bottled water is dominated by PA (polyamide) which is used in synthetic textile fabrics and clothing garments. The other polymers found in the MPs in beverages and bottled water are PEA and PET. Poly-ethylene-terephthalate (PET) is commonly used in plumbing (pipelines carrying water) and packaging like bottles, caps, containers, plastic bags, etc. Poly ester-amide (PEA) is yet another plastic (thermoplastic) used as an additive in plastics being used widely in rubber toughening, agriculture and food sectors. The presence of such polymers in the beverages and bottled water indicates the sources of MP contamination could vary from the source water used during the production process to the packaging material used. Use of properly treated water in the production of beverages and bottled water along with strict production and packaging protocols are the possible options to overcome or to reduce the MP contamination in such human consumable items.

REFERENCES

Aiswriya, V. P. (2019). *Microplastic Contamination in Carbonated Beverages- Soft Drinks and Soda.* Master degree (M.Sc.) dissertation, School of Environmental Sciences, Mahatma Gandhi University, Kottayam, Kerala, India, pp 1–51.

Cox, K. D., Covernton, G. A., Davies, H. L., Dower, J. F., Juanes, F., & Dudas, S. E., 2019. Human consumption of microplastics. *Environmental Science & Technology,* 53(12), 7068–7074.

Diaz-Basantes, M. F., Conesa, J. A., & Fullana, A., 2020. Microplastics in honey, beer, milk and refreshments in Ecuador as emerging contaminants. *Sustainability,* 12(14), 5514.

Ferraz, M., Bauer, A. L., Valiati, V. H., & Schulz, U. H., 2020. Microplastic concentrations in raw and drinking water in the Sinos River, Southern Brazil. *Water,* 12(11), 3115.

Ganesan, M., Nallathambi, G., & Srinivasalu, S., 2019. Fate and transport of microplastics from water sources. *Current Science,* 117(11), 00113891.

George, R. T., (2019). *Microplastics Contamination in Drinking Water Source of Kottayam District.* Master degree (M.Sc.) dissertation, School of Environmental Sciences, Mahatma Gandhi University, Kottayam, Kerala, India, pp 1–47.

Grumezescu, A. M., & Holban, A. M. eds., 2018. *Biopolymers for Food Design* (Vol. 20). Academic Press.

Johnson, A. C., Ball, H., Cross, R., Horton, A. A., Jurgens, M. D., Read, D. S., Vollertsen, J., & Svendsen, C., 2020. Identification and quantification of microplastics in potable water and their sources within water treatment works in England and Wales. *Environmental Science & Technology,* 54(19), 12326–12334.

Kankanige, D., & Babel, S., 2021. Contamination by≥ 6.5 µm-sized microplastics and their removability in a conventional water treatment plant (WTP) in Thailand. *Journal of Water Process Engineering, 40*, 101765.

Koelmans, A. A., Nor, N. H. M., Hermsen, E., Kooi, M., Mintenig, S. M., & De France, J., 2019. Microplastics in freshwaters and drinking water: Critical review and assessment of data quality. *Water Research, 155*, 410–422.

Kosuth, M., Mason, S. A., & Wattenberg, E. V. (2018). Anthropogenic contamination of tap water, beer, and sea salt. *PloS One, 13*(4), e0194970.

Kutralam-Muniasamy, G., Pérez-Guevara, F., Elizalde-Martínez, I., & Shruti, V. C., 2020. Branded milks–Are they immune from microplastics contamination?. *Science of the Total Environment, 714*, 136823.

Lachenmeier, D. W., Kocareva, J., Noack, D., & Kuballa, T., 2015. Microplastic identification in German beer-an artefact of laboratory contamination. *Deutsche Lebensmittel-Rundschau, 111*(10), 437–440.

Liebezeit, G., & Liebezeit, E., 2014. Synthetic particles as contaminants in German beers. *Food Additives & Contaminants: Part A, 31*(9), 1574–1578.

Mason, S. A., Welch, V. G., & Neratko, J., 2018. Synthetic polymer contamination in bottled water. *Frontiers in Chemistry, 6*, 407.

Mintenig, S. M., Löder, M. G. J., Primpke, S., & Gerdts, G., 2019. Low numbers of microplastics detected in drinking water from ground water sources. *Science of the Total Environment, 648*, 631–635.

Novotna, K., Cermakova, L., Pivokonska, L., Cajthaml, T., & Pivokonsky, M., 2019. Microplastics in drinking water treatment–Current knowledge and research needs. *Science of the Total Environment, 667*, 730–740.

Oßmann, B. E., Sarau, G., Holtmannspötter, H., Pischetsrieder, M., Christiansen, S. H., & Dicke, W., 2018. Small-sized microplastics and pigmented particles in bottled mineral water. *Water Research, 141*, 307–316.

Panno, S. V., Kelly, W. R., Scott, J., Zheng, W., McNeish, R. E., Holm, N., Hoellein, T. J., & Baranski, E. L., 2019. Microplastic contamination in karst groundwater systems. *Groundwater, 57*(2), 189–196.

Pivokonsky, M., Cermakova, L., Novotna, K., Peer, P., Cajthaml, T., & Janda, V., 2018. Occurrence of microplastics in raw and treated drinking water. *Science of the Total Environment, 643*, 1644–1651.

Prata, J. C., Paço, A., Reis, V., da Costa, J. P., Fernandes, A. J. S., da Costa, F. M., Duarte, A. C., & Rocha-Santos, T., 2020. Identification of microplastics in white wines capped with polyethylene stoppers using micro-Raman spectroscopy. *Food Chemistry, 331*, 127323.

Schymanski, D., Goldbeck, C., Humpf, H. U., & Fürst, P., 2018. Analysis of microplastics in water by micro-Raman spectroscopy: release of plastic particles from different packaging into mineral water. *Water Research, 129*, 154–162.

Sharma, S., & Chatterjee, S. (2017). Microplastic pollution, a threat to marine ecosystem and human health: a short review. *Environmental Science and Pollution Research, 24*(27), 21530–21547.

Shen, M., Zeng, Z., Wen, X., Ren, X., Zeng, G., Zhang, Y., & Xiao, R., 2021. Presence of microplastics in drinking water from freshwater sources: the investigation in Changsha, China. *Environmental Science and Pollution Research, 28*(31), 42313–42324.

Shruti, V. C., Pérez-Guevara, F., & Kutralam-Muniasamy, G., 2020a. Metro station free drinking water fountain-A potential "microplastics hotspot" for human consumption. *Environmental Pollution, 261,* 114227.

Shruti, V. C., Pérez-Guevara, F., Elizalde-Martínez, I., & Kutralam-Muniasamy, G., 2020b. First study of its kind on the microplastic contamination of soft drinks, cold tea and energy drinks-Future research and environmental considerations. *Science of the Total Environment, 726,* 138580.

Shruti, V. C., Pérez-Guevara, F., Elizalde-Martínez, I., & Kutralam-Muniasamy, G., 2021. Toward a unified framework for investigating micro (nano) plastics in packaged beverages intended for human consumption. *Environmental Pollution, 268,* 115811.

Sobhani, Z., Lei, Y., Tang, Y., Wu, L., Zhang, X., Naidu, R., Megharaj, M., & Fang, C., 2020. Microplastics generated when opening plastic packaging. *Scientific Reports, 10*(1), 1–7.

Strand, J., Feld, L., Murphy, F., Mackevica, A., & Hartmann, N. B., 2018. *Analysis of Microplastic Particles in Danish Drinking Water* (p. 34). DCE-Danish Centre for Environment and Energy.

Tong, H., Jiang, Q., Hu, X., & Zhong, X., 2020. Occurrence and identification of microplastics in tap water from China. *Chemosphere, 252,* 126493.

Toussaint, B., Raffael, B., Angers-Loustau, A., Gilliland, D., Kestens, V., Petrillo, M., Rio-Echevarria, I. M., & Van den Eede, G., 2019. Review of micro-and nanoplastic contamination in the food chain. *Food Additives & Contaminants: Part A, 36*(5), 639–673.

Uhl, W., Eftekhardadkhah, M., & Svendsen, C., 2018. Mapping microplastic in Norwegian drinking water. *Atlantic, 185,* 491–497.

Welle, F., & Franz, R., 2018. Microplastic in bottled natural mineral water–literature review and considerations on exposure and risk assessment. *Food Additives & Contaminants: Part A, 35*(12), 2482–2492.

Wiesheu, A. C., Anger, P. M., Baumann, T., Niessner, R., & Ivleva, N. P., 2016. Raman microspectroscopic analysis of fibers in beverages. *Analytical Methods, 8*(28), 5722–5725.

Wilkinson, J., Hooda, P. S., Barker, J., Barton, S., & Swinden, J., 2017. Occurrence, fate and transformation of emerging contaminants in water: An overarching review of the field. *Environmental Pollution, 231,* 954–970.

Zhang, M., Li, J., Ding, H., Ding, J., Jiang, F., Ding, N. X., & Sun, C., 2020. Distribution characteristics and influencing factors of microplastics in urban tap water and water sources in Qingdao, China. *Analytical Letters, 53*(8), 1312–1327.

Zhou, X. J., Wang, J., Li, H. Y., Zhang, H. M., & Zhang, D. L., 2021. Microplastic pollution of bottled water in China. *Journal of Water Process Engineering, 40,* 101884.

Zuccarello, P., Ferrante, M., Cristaldi, A., Copat, C., Grasso, A., Sangregorio, D., ..., & Conti, G. O., (2019). Exposure to microplastics (< 10 μm) associated to plastic bottles mineral water consumption: The first quantitative study. *Water Research, 157,* 365–371.

Methods in Microplastics Extraction and Identification

4

4.1 INTRODUCTION

The production of plastic has increased severalfold during the past five decades and reached nearly 370 million tons per year (Plastic Europe, 2019). The increased global consumption of plastics has also led to an increase in plastic waste; only 9–21% of plastic waste is either recycled or incinerated while the rest goes to landfills or illegal dumping/disposal (UNEP, 2019). As a result of the mismanagement of plastic waste, 8–12 million tons of annual production of plastic (IUCN, 2018) end up in oceans every year through various outlets including rivers (IUCN, 2018). This plastic debris either floats on the sea surface or accumulates in deep-sea sediments and shorelines (IUCN, 2018; Suaria and Aliani, 2014; Eriksen et al., 2014; Browne et al., 2011), causing harm to marine biota and becoming a severe threat to seafood quality and safety when this plastic debris fragments into tiny pieces of microplastics (plastic particles < 5 mm in size) and is consumed by the marine/freshwater biota as their feed.

A wealth of literature exists on microplastic contamination in marine as well as freshwater biota including fishes and other seafood items for human consumption. However, different techniques adopted in sampling and analytical procedures by the researchers pose an issue of comparison and repeatability of the findings reported. There is also variation in the units of reporting the abundance of MPs, such as the number of MPs per kg of sediment *or* per m^2/km^2 of the surface area of the sediment; in the case of water, the number of MPs per liter/m^3 *or* per m^2/km^2 of surface area trawled (sampled); and in the case of biota, the number of MPs per number of individuals *or* per g of edible tissue, etc. This leads to difficulty in comparison of the results. In the current

DOI: 10.1201/9781003201755-4

chapter, information available in the literature on the extraction protocol of MPs, digestion of organic matter, assessment of morphological characteristics and polymer identification are compiled to provide the reader an overview of methodologies being adopted in MPs research.

The sampling protocol for MPs study varies depending upon the category of samples such as water, sediment, soil, biota and air; most of the sampling techniques for each of these categories have been standardized globally. However, the methods adopted in the extraction of MPs from the sample and polymer identification procedures differ considerably (Table 4.1). The objective of this chapter is to summarize various techniques reported in the literature for the extraction and polymer identification of MPs from different samples.

4.2 MICROPLASTICS EXTRACTION

Extraction of MPs from the sample is a crucial step leading to the determination of the abundance, physical and chemical characteristics of MPs. The most common protocol followed by many researchers for the extraction of MPs is the method described by the National Oceanic and Atmospheric Administration (NOAA) of the USA (Masura et al., 2015). This protocol was mainly developed to study MP contamination in the marine environment and also serves as the basis for the extraction of MPs from samples having an organic rich matrix with suitable modifications.

To extract the MPs from any kind of sample, the initial step would be sieving with a metal sieve of appropriate pore size (in order to remove particles of larger dimensions, usually > 5 mm) followed by digestion of organic matter present in the sample, density separation of the digested sample in order to make the MPs to float on the surface and filtration of the supernatant (the portion having most of the MPs) over a suitable filter in order to retain the MPs over the filter paper for subsequent observation and polymer analysis. These steps are presented in Figure 4.1. The variations in this procedure generally noted in the literature are at the digestion step, in which the reagents used may vary depending on the sample type and its organic content; the reagents and their concentration used in the preparation of density separation medium also vary and the filter paper used may be either glass fiber (GF) of different grades or nitro-cellulose. The details of each step of the extraction procedure are explained below.

4.2.1 Initial Sieving of Sample

As the dimensions of microplastics are < 5 mm, the samples – especially water samples – will be initially screened through a metallic screen with a pore size

TABLE 4.1   Digestion Techniques Adopted in Extraction of MPs from Human Food Items

HUMAN FOOD ITEMS	DIGESTION METHOD	REFERENCES
Fishes	Alkaline	Sathish et al., 2020; Karuppasamy et al., 2020; Goswami et al., 2020; Devi et al., 2020; Karbalaei et al., 2019; Nie et al., 2019; Al-Lihaibi et al., 2019; Fang et al., 2019; Bessa et al., 2018; Morgana et al., 2018; Kumar et al., 2018; Karami et al., 2017; Wang et al., 2017; Tanaka and Takada, 2016; Rochman, 2015; Lusher et al., 2016
	Acid	Zhu et al., 2019; Chan et al., 2019; Lu et al., 2016
	Oxidation	James et al., 2020; Li et al., 2020; Hossain et al., 2019; Blettler et al., 2019; Su et al., 2019; Lv et al., 2019; Zhang et al., 2019; Digka et al., 2018; Avio et al., 2015
	Enzymatic	Pozo et al., 2019
	Combination	Al-Salem et al., 2020; Zhang et al., 2019; Yuan et al., 2019; Slootmaekers et al., 2019; Collard et al., 2017
Bivalves/ mussels	Alkaline	Phuong et al 2018a & b; Claessens et al., 2013; Van Cauwenberghe et al., 2014, Van Cauwenberghe et al., 2013; Davidson and Dudas, 2016;
	Acid	Van Cauwenberghe and Janssen, 2014; Van Cauwenberghe et al., 2015; Murphy, 2018; Vandrmeersch et al., 2015b; Santana et al., 2016
	Oxidation	Mathalon and Hill, 2014; Li et al., 2018, 2016, 2015; Kolandhasamy et al., 2018; Qu et al., 2018; Bonello et al., 2018; Digka et al., 2018a & b; Khoironi and Anggoro, 2018;
	Enzymatic	Courtene-Jones et al., 2017; Catarino et al., 2018, 2017; Karlsson et al., 2017
	Combination	De Witte et al., 2014; Leslie et al., 2017; Vandrmeersch et al., 2015b; Devriese et al., 2015
Drinking water	Oxidation	Anderson et al., 2017; Mason et al., 2018; Schymanski et al., 2018; Oßmann et al., 2018; Pivokonsky et al., 2018; Koelmans et al., 2019
Salts/ honey	Oxidation	Yang et al., 2015; Liebezeit and Liebezeit, 2013, 2015; Gündoğdu, 2018; Yang et al., 2018; Seth and Shriwastav, 2018; Zhang et al., 2019; Sathish et al., 2020
	Alkaline	Enders et al., 2017; Karami et al., 2018; Lusher et al., 2020

of 5 mm (in order to remove particles of more than 5 mm size); particles that pass through a 5 mm sieve will be subsequently sieved with another (metallic) sieve with lower pore size, maybe 300 or 100 or 50 μm. The particles retained over the second sieve will be collected and processed further. In summary, this kind of sieving with a set of sieves having a higher pore size on top and lower pore size at the bottom will result in obtaining particles of size ranging between the pore sizes of the sieves. The particles retained over the sieve (with lower pore size) should be rinsed using ultrapure water and transferred to glass beakers or storage containers for further analysis (Viršek et al., 2016; Cutroneo et al., 2020; Dubaish and Liebezeit, 2013; Desforges et al., 2014). This kind of sieving of the samples is more applicable to liquid samples and to some extent sediment samples but not suitable for biotic samples. This preliminary screening helps in reducing the volume of water samples to be transported from the field to the laboratory for further analysis.

4.2.2 Digestion of Sample

Sample digestion aims to remove the organic particles present in the sample without affecting the quantity or quality of the MPs in the sample. In a few cases, the digestion step is avoided as the presence of organic matter is quite negligible or absent, as in the case of bottled water. The common digestion methods adopted are acid digestion (Zobkov et al., 2019), alkaline digestion (Zhu et al., 2019), enzymatic digestion (Saliu et al., 2018) and oxidation (Pan et al., 2019). In some cases, a combination of two digestion techniques may be adopted in sequence (Tamminga et al., 2018). The details of digestion methods used in human food items are summarized in Table 4.1.

4.2.2.1 Acid Digestion

Acid digestion techniques which are common in trace metal analyses have been introduced in microplastic research (Miller et al., 2017). This technique is found to be more appropriate for the degradation of organic matter of animal tissues or organs at certain temperature (Qiu et al., 2016). Commonly used acids in digestion techniques are HCl, HNO_3, $HClO_4$ and HClO with a combination of H_2O_2 or without it. Among these, HNO_3 is widely used (Toussaint et al., 2019; Zhu et al., 2019; Enders et al., 2017); however, it has been observed by researchers that in a few cases digestion with HNO_3 has resulted in the loss of polymers which have low resistance to acids like nylon; melting of PE, PP and yellowing of polymers such as PP, PVC, etc. (Catarino et al., 2017; Karami et al., 2017b; Maes et al., 2017). It is also noticed that strong acids, such as HCl or HF, can affect the morphological integrity of the microplastic particles,

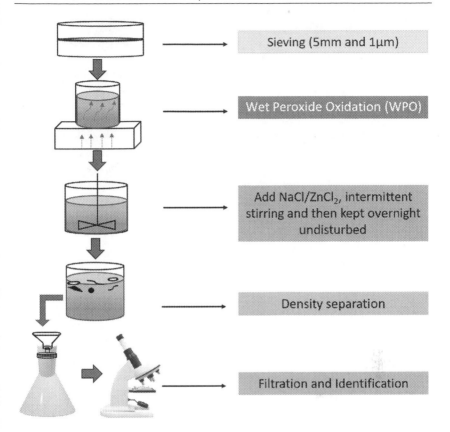

FIGURE 4.1 Flowchart of microplastic analysis.

hence the use of such strong acids is not recommended in MPs extraction studies. (Cutroneo et al., 2020). In summary, the use of acids for digestion may be adopted with the precaution of avoiding strong acids at high concentrations and heating the digestion mixture above 60° C, etc. (Maes et al., 2017; Munno et al., 2018).

4.2.2.2 Alkaline Digestion

Alkaline digestion is similar to acid digestion and widely used in the degradation of organic matter, especially while extracting MPs from biotic samples (Sathish et al., 2020; Karuppasamy et al., 2020; Goswami et al., 2020; Al-Lihaibi et al., 2019; Fang et al., 2019; Bessa et al., 2018; Morgana et al., 2018; Kumar et al., 2018; Karami et al., 2017; Wang et al., 2017; Tanaka and Takada, 2016; Rochman, 2015; Lusher et al., 2016; Davidson and Dudas, 2016;

Van Cauwenberghe and Janssen, 2014, Van Cauwenberghe et al., 2013). NaOH and KOH are the common alkalis used as digestive reagents. Among these, KOH has good digestion potential for organic matter (Catarino et al. 2017; Munno et al., 2018). However, damage caused to a few polymers has also been reported due to the use of KOH as a saturated solution (1,120 g/L of H_2O) and the heating temperature goes beyond 50° C. Also, at this concentration, KOH digestion may lead to spectra deviations and lower-quality Raman spectra of the resulting polymers when compared to undigested polymers (Martinelli et al., 2020; Enders et al., 2017; Karami et al., 2017). A moderate concentration of KOH solution (10%) was found to be effective in removing 97.1–98.9% of organic matter from the sample consisting of ground fish tissue at temperatures ranging from 25 to 50° C (Karami et al., 2017). Lusher et al. (2020) recommended KOH as more effective for the digestion of biotic samples with a combination of other extraction methods.

The limitations of using NaOH as a digestion solution have also been reported in the literature, like degradation of the plastic particles such as PA and PET and alters the color of PVC and PET (Dehaut et al., 2016). Thus, irrespective of NaOH or KOH, the use of moderate concentration and maintaining the digestion temperature below 50° C would provide efficient removal of organic matter without affecting the quality or abundance of the MPs in the sample.

A comparative study by Enders et al. (2016) has indicated alkaline digestion as a better choice over acid digestion as alkali digestion is applicable to a wider range of polymers with a relatively lower risk of damage to the particles in terms of chemical as well as structural modifications (Toussaint et al., 2019).

4.2.2.3 Oxidative Agents

Hydrogen peroxide (H_2O_2) is an efficient oxidative agent used in the degradation of organic matter in most samples. As per the recommendation of the National Oceanic and Atmospheric Administration (NOAA), the use of H_2O_2 (30%) with iron sulfate (Fenton's reagent) in wet oxidation of the samples by heating at 75° C using a glass beaker (Masura et al., 2015) is suitable to both sediment and water samples. H_2O_2 has been reported to digest organic matter more efficiently than alkali and acid digestion techniques, more importantly with the least or no damage to the MPs (Mae et al., 2017; Nuelle et al., 2014; Qiu et al., 2016; Zhao et al., 2017). The temperature used in heating the oxidation mixture plays a prominent role as reported by many researchers: H_2O_2 oxidation done at ambient temperature for seven days has resulted only in the removal of 25% of organic matter (Cole et al., 2014), while the use of H_2O_2 (15%) at 50° C has removed organic matter efficiently in an overnight period (Avio et al., 2020). Oxidation with H_2O_2 or as Fenton's reagent (H_2O_2 with iron sulfate catalyst) with modifications in the heating temperature (50–60° C) is predominantly found in the literature (Pfohl et al., 2021; Prata et al., 2019; Hurley et al., 2018; Tagg et al., 2017; Sruthy and Ramasamy, 2017; Flotron et al., 2005) and is often referred to as wet peroxidase oxidation (WPO).

4.2.2.4 Enzymatic Digestion

Enzymatic digestion of samples for MPs extraction was introduced for the reason that this technique is less hazardous with no impact on the polymers (Maes et al., 2017). However, due to the biological specificity of the enzymes, this method is target specific, thus its usage was noted with limitations; for example, the inaction of enzymes on non-target organic matter has led to the use of a combination of enzymes which has hiked the cost. Proteinase-K, trypsin, collagenase, papain80 and commercially isolated pancreatic enzymes (PEz) are the enzymes used by several researchers for the digestion of samples for the extraction of MPs (Prata et al., 2019; Karlsson et al., 2017; Courtene-Jones et al., 2017; Loder et al., 2017; Catarino et al., 2017).

Except for proteinase-K, the rest of the enzymes were observed to have an efficiency of around 86% while proteinase-K was noted as having 88% efficiency in the digestion of organic matter (Friesen et al., 2019). There are studies that report on the attempts with a combination of enzymes (Friesen et al., 2019; Railo et al., 2018; Karami et al., 2017), multiple phases of digestion like enzymatic digestion followed by H_2O_2 and again with enzymes; these attempts have achieved around 97–98% efficiency in the digestion of organic matter and took 16 days for the completion of this protocol (Lusher et al., 2020). Overall review of enzymatic digestion indicates that they are being used on a small scale mainly due to the higher cost and the complex digestion procedure involving multiple steps using the chemicals along with a combination of enzymes (Loder et al., 2017; Toussaint et al., 2019).

4.2.2.5 Sequential Digestion

Sequential digestion involves the use of multiple digestion reagents in a sequence in order to get a better removal of organic matter which is essential in extracting MPs from complex samples. Roch and Brinker (2017) have used acid followed by alkali digestion of biological material for the better recovery of MPs. Caution is required when such multiple digestion reagents are used, as such reagents may affect the polymer; for instance, the International Council for the Exploration of the Sea (ICES, 2015) has recommended nitric acid (69%) and perchloric acid (70% $HClO_4$) but this was noted to have caused damages to common polymers like polyamide (PA) and polyurethane (PU) (Enders et al., 2017; Lusher et al., 2020). Lusher et al. (2020) found the combination of KOH and NaClO (sodium hypochlorite) in a 1:1 ratio more effective in the digestion of fish tissue than KOH alone.

In summary, alkaline and acid digestion techniques are found to be more applicable to solid non-soluble animal tissues and organs, while oxidative agents are appropriate for liquid and water-soluble food items. Enzymatic digestion methods are suitable for many biological matrices; however, these techniques are reported as expensive, and on many occasions, for complete digestion, a combination of several enzymes and other digestion reagents is needed.

4.2.3 Density Separation

After complete digestion of organic matter, the next step is density separation in which the MPs are made to float to the surface of the denser medium. Generally, floatation (density separation) is done by adding density solution – prepared by the addition of an appropriate chemical of higher density to water – to the container having the digested mixture. Proper mixing of the sample is necessary after the addition of the density solution in order to ensure that MPs detach from the sample matrix, which is achieved by various techniques including centrifugation, rotary shaking, magnetic stirring (continuous or intermittent) and periodic manual stirring using glass rods, etc. Tall glass beakers are commonly used for this purpose; however, the use of centrifuge tubes and separating funnels is also reported in the literature (Hurley et al., 2018; Scheurer and Bigalke, 2018; Besley et al., 2017; Woodall et al., 2014). Duration and type of mixing vary depending on the sample; similarly, the time required to complete the density separation (floatation time) also varies from a couple of minutes to 24 hours (Hurley et al., 2018; Karlsson et al., 2017; Thompson et al., 2004).

Preparation of a density solution with appropriate chemicals is a critical step in floatation. Sodium chloride (NaCl) is one of the commonly used chemicals adopted by many researchers and recommended by many agencies including the National Oceanic and Atmospheric Administration (NOAA) as it is less expensive and harmless to the user and the environment. However, NaCl can result in a density solution with a density range of 1.0 to 1.2 (g/cm^3); as much denser solutions are needed to separate polymers of higher densities, there are a few more chemicals used for this purpose as listed in Table 4.2.

One among the important criteria in the selection of these chemicals besides cost and resulting density range of the solution is the toxic/hazardous nature of the chemical and the recyclability of the used solution. In this regard, $ZnCl_2$ is much used in the extraction of MPs from sediment samples, resulting in a density range of 1.6–1.8 (g/cm^3); it has a high recovery rate of MPs and

TABLE 4.2 Chemicals Used in Preparation of Density Solution

NAME OF CHEMICAL	CHEMICAL FORMULA	DENSITY (G/CM³)	LEVEL OF TOXICITY
Sodium chloride	NaCl	1.0–1.2	Low
Sodium bromide	NaBr	1.37–1.40	Low
Sodium tungstate dehydrate	$Na_2WO_4 \cdot 2H_2O$	1.40	Low
Sodium polytungstate	$3Na_2WO_4 \cdot 9WO_3 \cdot H_2O$	1.40	Low
Lithium metatungstate	$Li_6(H2W_{12}O_40)$	1.6	Moderate
Sodium iodide	NaI	1.80	Moderate
Zinc bromide	$ZnBr_2$	1.71	High
Zinc chloride	$ZnCl_2$	1.60–1.80	High

Source: Baseman (2018).

is less expensive but it is extremely hazardous and corrosive (Coppock et al., 2017; Zobkov et al., 2017; Imhof et al., 2012). Sodium iodide (NaI) is another appropriate chemical resulting in a density of 1.8 (g/cm³) with moderate toxicity and is also recyclable, but it is expensive.

4.2.4 Filtration

Filtration is the last step – after screening, digestion and density separation – in the sequence of sample processing for the extraction of MPs. This is an important step involving the transfer of microplastic particles from the density solution over a filter paper with the help of a filtration unit attached to a vacuum pump. The importance attached to this step is due to the fact that all further assessment of MPs from the sample in terms of both quantitative and qualitative measures depends on the MPs retained over the filter paper.

There are different types of filtration units used by researchers, starting with a polysulfone filter funnel connected with a vacuum pump, a three-place manifold and a Buchner flask attached to a vacuum pump; among these, the Buchner flask unit is commonly used in many studies (Figure 4.1). The filter paper used in the filtration unit also varies as per the need; Whatman glass fiber (GF) filter paper grade GF/A, GF/C or GF/F is the most commonly used filter paper, while nitrocellulose filter paper and isopore filters are also used

(Hanvey et al., 2017). Recently, a few studies have used Anodisc filters (alumina oxide) which are useful in further analysis using micro-FT IR (Primpke et al., 2020). Once filtration is over, the filter paper is subsequently dried in a glass petri dish at 40° C in an oven prior to subjecting the filter paper along with the particles retained over it for further analysis such as morphological characterization (shape, size and color) and polymer identification.

4.3 MICROSCOPIC TECHNIQUES USED IN ASSESSING THE MORPHOTYPE, COLOR, SIZE AND ABUNDANCE OF MPS

Visual observation of the MPs present over the filter papers is the next step. The filter papers after drying are observed under a microscope to study the shape, size, color and number of MPs retained over that filter paper. MPs of larger size (particle size > 5 mm) can be visually identified with the naked eye. Different types of microscopes like stereo, optical and dissection microscopes have been used for this purpose. A brief description of popular techniques used in assessing the characteristics of MPs is provided in the subsequent paragraphs, while the advantages and limitations of each technique are summarized in Table 4.3.

4.3.1 Stereomicroscope, Optical and Dissection Microscope

The light beam of a stereomicroscope comes from above while in an optical microscope light is from below and penetrates through the object. Thus in a stereomicroscope, three-dimensional analysis by observing the sample from two slightly different angles is possible. At higher magnification, the details of the surface structure of the MPs could be studied. The stereomicroscope is widely used in studying MPs extracted from biota, water, salt and other food items (Prata et al., 2020; Karuppasamy et al., 2020; James et al., 2020; Fang et al., 2019; Karami et al., 2017) while the optical microscope was used by Mason et al. (2018) (water sample) and the dissection microscope by Sathish et al. (2020), Kosuth et al. (2018) and Liebezeit and Liebezeit (2013) for MPs from honey, sugar and beer. Microscope magnification used in MP studies varies according to the sample and MP size. Stereo microscopes with calibration scales and camera attachments are used to photograph the images along with

TABLE 4.3   Visual Observation Techniques for Microplastics: Advantages and Limitations

TECHNIQUES		ADVANTAGES	LIMITATIONS
Morphological identification	Optical/stereo microscope	• Easy to handle and use • Low cost • Able to identify MPs' shape, size and color	• Unable to provide information on plastic composition or polymer type
	Fluorescence microscope	• Able to identify the luminance particle of plastics • Recognizes the shape and size of the particles	• Expensive • Allows only to observe the specific structure, which is labeled for fluorescence
	SEM/EDX	• Provides higher-resolution images of particles • Able to find MPs' shape and elements compositions	• Expensive • Specially trained manpower is needed to operate the instrument • Requires special sample preparation like conductive coating, drying and cleaning
Polymer identification	FT IR/micro-FT IR	• Identifies the polymer spectrum of the MPs • Able to identify small MP particles (~20 μm)	• Expensive • Time consuming and tedious • Difficult to identify the particles smaller than 20 μm
	Raman/ micro-Raman spectroscopy	• Identifies the polymer spectrum of the particles • Non-destructive technique • Able to recognize small MP particles (1 μm) and nanoplastic (< 1 μm) particles too	• Expensive instrument • Specially trained man power is needed to operate the instrument • Unable to recognize the luminance particle of plastics • Time consuming

(Continued)

TABLE 4.3 (CONTINUED)   Visual Observation Techniques for Microplastics: Advantages and Limitations

TECHNIQUES		ADVANTAGES	LIMITATIONS
Other techniques	Thermal analysis (Py-GC-MS)	• Simple and fast • Able to identify the polymers and other organic additives	• Cost effective • Requires expert person to operate • Unable to determine shape, size and numbers • Destructive technique
	Fluorescent tagging with Nile red	• Able to detect and quantify particles between 1 mm –20 μm • Time saving and semi-automated	• Unable to identify the polymer type • Expensive
	Time-of-flight secondary ion mass spectrometry (ToF-SIMS)	• Provides a high-resolution image • Suitable for mixture of particles	• Very expensive • Sample must be vacuum compatible

the size of the MPs observed (Ramasamy et al., 2019); a few such images of MPs are shown in Figure 4.1.

4.3.2 Fluorescence Microscope

In contrast to the optical microscope in which the image is viewed by the reflection of the light on the sample, in a fluorescence microscope, the fluorescent emission from the sample that is excited by a specific wavelength is collected and viewed. Apart from studying biological samples (cells, bacteria), the fluorescence microscope is also used to study MPs (Prudent and Raoult, 2019), especially for viewing white and transparent MPs on the basis of their innate ability to emit fluorescence (Noren, 2008). This microscope has been used in a few studies for the identification of MPs in food items (Dowarah et al., 2020; Shruti et al., 2020). A fluorescent dye (Nile red) is used to label MPs in the samples and is identified based on the emissions using a fluorescence microscope (Cole et al., 2013).

4.3.3 Scanning Electron Microscope (SEM) and Energy Dispersive X-Ray (EDX)

A few studies have used scanning electron microscope (SEM) to visualize the surface characteristics and shape of MPs found in human food items (Karami et al., 2017; Shruti et al., 2020; Karbalaei et al., 2019; Karuppasamy et al., 2020; Sathish et al., 2020). This microscope provides a high-resolution image of MPs which may provide details of the cracks present on the surface of MPs indicating the possibility of further fragmentation so that the bioavailability of the MPs increases, which in turn enhances the chance of trophic-level transfer of MPs. When coupled with energy dispersive X-ray (EDX) it can provide the elemental composition of the sample (Naeeji et al., 2020; Oßmann et al., 2019; Gniadeka and Dąbrowskab, 2019; Shruti et al., 2019; Sathish et al., 2020).

4.4 POLYMER PROFILE IDENTIFICATION TECHNIQUES

After studying the physical features of the MPs extracted from the samples, identification of the polymer profile of the MPs is the next crucial step. In this step, sophisticated spectroscopic instruments and spectral identification library (software) are to be used. A summary of techniques employed in the identification of the polymer profile of MPs is given in Table 4.3. A brief description of those techniques is also discussed below

4.4.1 FT IR/Micro-FT IR

Fourier-transform infrared (FTIR) spectroscopy is widely used to identify the MPs in human consumable items and compositions of the polymer. The attenuated total reflection FTIR (ATR-FTIR) can identify the MPs of more than 500 μm in size (Jung et al., 2018). The ATR-FTIR instrument when coupled with a microscope, then known as micro-FTIR (μ-FTIR), can detect the smaller MP particles of ~20 μm (Elkhatib and Oyanedel-Craver, 2020; Toussaint et al., 2019). Micro-FTIR could detect particles of such smaller size as it is attached to a microscope and its function is based on the interaction between infrared radiation and matter (Käppler et al., 2015). The spectrum generated by the

FTIR/micro-FTIR can be analyzed using the IR spectrum database library in order to identify the polymer content. The advantage of this technique is it is a non-destructive technique and requires a small quantity of the sample. The major limitations are: it is expensive, time consuming and cannot identify particles of smaller than 20 μm size.

4.4.2 Raman Spectroscopy/ Micro-Raman Spectroscopy

Raman spectroscopy is an advanced technique with the ability to detect MPs of less than 20 μm and up to 1 μm size. This technique has been widely used in studying MPs from human food items (Shruti et al., 2020; Prata et al., 2020b; Al-Salem et al., 2020; Karbalaei et al., 2019; Nie et al., 2019; Yuan et al., 2019; Karami et al., 2017; Dowarah et al., 2020; Oßmann et al., 2018; Schymanski et al., 2018). Raman spectroscopy provides images of higher resolution than IR spectroscopy as it is a laser-based method. Similar to FTIR, Raman spectroscopy also generates the spectrum which can be analyzed using a spectral library to identify the polymer type of MPs. The advantage of this technique is, it provides high-resolution images and particles up to 1 μm in size can also be studied. The major limitations are: unable to identify auto-fluorescence particles and it is expensive.

4.4.3 Other Techniques

There are a few other techniques that are also used in the identification and characterization of MP particles.

4.4.3.1 Thermal Analysis: Pyrolysis GC-MS Method

In this technique, pyrolysis and gas chromatography are combined with the MS detector (Py-GC-MS). In this method, MP particles are subjected to thermal degradation and the resulting products are analyzed to identify polymer types and the organic additives present in the MPs. As no solvents are used, no interference of background contamination is observed in this method. Usually, this is done in one single run without the use of solvents, thus avoiding background contamination. Major limitations of this technique are: it is not possible to assess the number, size or shape of the particles; and as it is a destructive technique, other than mass concentration, no other results pertaining to the morphotype of MPs are possible (Conesa and Iñiguez, 2020).

4.4.3.2 Time-of-Flight Secondary Ion Mass Spectrometry (ToF-SIMS)

This technique results in high spatial-resolution images and is suitable for a mixture of particles; however, the technique is complex and quite expensive. The sample used in this technique must be vacuum compatible as weathering of the sample may complicate the identification. This technique has been used in analyzing polyethylene (PE) MPs in marine water (Jungnickel et al., 2016).

4.4.4 Spectral Analysis for Polymer Identification

As mentioned earlier, the spectroscopic techniques result in spectra as their output, corresponding to the polymer type of the MPs. To analyze these spectra for identifying the polymer type, various databases or polymer libraries are available; all the resulting spectra are to be compared with the reference spectra provided in the libraries. The more commonly used polymer identification libraries/databases/software are the KnowItAll software, OMNIC software, OPUS software, Bruker database, rapID-reference library, Berlin and the Commercial Spectral Library. All the obtained spectra can be run through any of these polymer library–based software to determine the type of polymer. A flow diagram of all the extraction and polymer identification is given in Figure 4.2.

4.5 CONCLUSION

The global publications based on microplastics (MPs) are increasing at a faster rate compared to a decade ago. However, there are a few limitations that need to be addressed in order to make the findings of these publications more useful. The limitations pertain to the lack of uniformity in the methods/protocol being adopted in sampling, extraction of MPs from environmental matrices, the procedures of analysis and reporting of findings. As far as the MP contamination of human food items are concerned, the limitations are more visible as this particular field of research is in its nascent stage. The major impact of these dissimilarities in the protocols leads to restrictions in the comparison of results being published globally. The evolution of analytical techniques and instruments being used in recent times also imposes more restrictions on

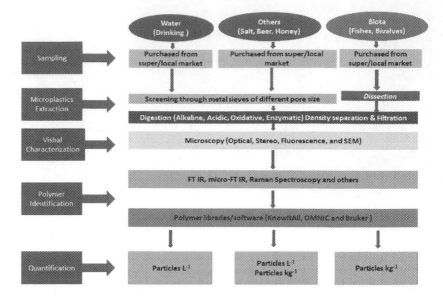

FIGURE 4.2 The summary of protocol being followed in extraction and analysis of microplastic particles from human consumable items.

the comparison of results published in yesteryears as some of the methods adopted become outdated. In this fast-growing field of microplastics analysis, the development of standard methods with uniformity is the global need of the hour. Except for the *Laboratory Methods for the Analysis of Microplastics in the Marine Environment* introduced by the National Oceanic and Atmospheric Administration (NOAA) in 2015 (Masura et al., 2015), no other manual has been released so far. Thus, it is high time that global organizations like the WHO, USEPA, etc. should come forward to standardize the sampling and analytical procedures being adopted in MPs research for uniformity and for ease of comparison of the findings across the globe.

REFERENCES

Al-Lihaibi, S., Al-Mehmadi, A., Alarif, W. M., Bawakid, N. O., Kallenborn, R., & Ali, A. M. (2019). Microplastics in sediments and fish from the Red Sea coast at Jeddah (Saudi Arabia). *Environmental Chemistry*, *16*(8), 641–650.

Al-Salem, S. M. M., Uddin, S., & Lyons, B. (2020). Evidence of microplastics (MP) in gut content of major consumed marine fish species in the State of Kuwait (of the Arabian/Persian Gulf). *Marine Pollution Bulletin*, *154*, 111052.

Anderson, P. J., Warrack, S., Langen, V., Challis, J. K., Hanson, M. L., & Rennie, M. D. (2017). Microplastic contamination in lake Winnipeg, Canada. *Environmental Pollution*, *225*, 223–231.

Avio, C. G., Gorbi, S., Milan, M., Benedetti, M., Fattorini, D., d'Errico, G., ..., & Regoli, F. (2015). Pollutants bioavailability and toxicological risk from microplastics to marine mussels. *Environmental Pollution*, *198*, 211–222.

Avio, C. G., Pittura, L., d'Errico, G., Abel, S., Amorello, S., Marino, G., Gorbi, S., & Regoli, F. (2020). Distribution and characterization of microplastic particles and textile microfibers in Adriatic food webs: General insights for biomonitoring strategies. *Environmental Pollution*, *258*, 113766.

Bessa, F., Barría, P., Neto, J. M., Frias, J. P., Otero, V., Sobral, P., & Marques, J. C. (2018). Occurrence of microplastics in commercial fish from a natural estuarine environment. *Marine Pollution Bulletin*, *128*, 575–584.

Besley, A., Vijver, M. G., Behrens, P., & Bosker, T. (2017). A standardized method for sampling and extraction methods for quantifying microplastics in beach sand. *Marine Pollution Bulletin*, *114*(1), 77–83.

Blettler, M. C., Garello, N., Ginon, L., Abrial, E., Espinola, L. A., &Wantzen, K. M. (2019). Massive plastic pollution in a mega-river of a developing country: Sediment deposition and ingestion by fish (Prochilodus lineatus). *Environmental Pollution*, *255*, 113348.

Bonello, G., Varrella, P., & Pane, L. (2018). First evaluation of microplastic content in benthic filter-feeders of the Gulf of La Spezia (Ligurian Sea). *Journal of aquatic food product technology*, *27*(3), 284–291

Browne, M. A., Crump, P., Niven, S. J., Teuten, E., Tonkin, A., Galloway, T., & Thompson, R. (2011). Accumulation of microplastic on shorelines worldwide: Sources and sinks. *Environmental Science & Technology*, *45*(21), 9175–9179.

Catarino, A. I., Macchia, V., Sanderson, W. G., Thompson, R. C., & Henry, T. B. (2018). Low levels of microplastics (MP) in wild mussels indicate that MP ingestion by humans is minimal compared to exposure via household fibres fallout during a meal. *Environmental Pollution*, *237*, 675–684.

Catarino, A. I., Thompson, R., Sanderson, W., & Henry, T. B. (2017). Development and optimization of a standard method for extraction of microplastics in mussels by enzyme digestion of soft tissues. *Environmental Toxicology and Chemistry*, *36*(4), 947–951.

Chan, H. S. H., Dingle, C., & Not, C. (2019). Evidence for non-selective ingestion of microplastic in demersal fish. *Marine Pollution Bulletin*, *149*, 110523.

Claessens, M., Van Cauwenberghe, L., Vandegehuchte, M. B., & Janssen, C. R. (2013). New techniques for the detection of microplastics in sediments and field collected organisms. *Marine Pollution Bulletin*, *70*(1–2), 227–233.

Cole, M., Lindeque, P., Fileman, E., Halsband, C., Goodhead, R., Moger, J., & Galloway, T. S. (2013). Microplastic ingestion by zooplankton. *Environmental Science & Technology*, *47*(12), 6646–6655.

Cole, M., Webb, H., Lindeque, P. K., Fileman, E. S., Halsband, C., & Galloway, T. S. (2014). Isolation of microplastics in biota-rich seawater samples and marine organisms. *Scientific Reports*, *4*(1), 1–8.

Collard, F., Gilbert, B., Compère, P., Eppe, G., Das, K., Jauniaux, T., & Parmentier, E. (2017). Microplastics in livers of European anchovies (Engraulis encrasicolus, L.). *Environmental Pollution*, *229*, 1000–1005.

Conesa, J. A., & Iñiguez, M. E. (2020). Analysis of microplastics in food samples. In *Handbook of Microplastics in the Environment* by Rocha-Santos, T., Costa, M., & Mouneyrac, C. (Eds.). Springer International Publishing.

Coppock, R. L., Cole, M., Lindeque, P. K., Queirós, A. M., & Galloway, T. S. (2017). A small-scale, portable method for extracting microplastics from marine sediments. *Environmental Pollution*, *230*, 829–837.

Courtene-Jones, W., Quinn, B., Gary, S. F., Mogg, A. O., & Narayanaswamy, B. E. (2017). Microplastic pollution identified in deep-sea water and ingested by benthic invertebrates in the Rockall Trough, North Atlantic Ocean. *Environmental Pollution*, *231*, 271–280.

Cutroneo, L., Reboa, A., Besio, G., Borgogno, F., Canesi, L., Canuto, S., Dara, M., Enrile, F., Forioso, I., Greco, G., & Lenoble, V. (2020). Microplastics in seawater: Sampling strategies, laboratory methodologies, and identification techniques applied to port environment. *Environmental Science and Pollution Research*, *27*(9), 8938–8952.

Davidson, K., & Dudas, S. E. (2016). Microplastic ingestion by wild and cultured Manila clams (Venerupis philippinarum) from Baynes Sound, British Columbia. *Archives of Environmental Contamination and Toxicology*, *71*(2), 147–156.

De Witte, B., Devriese, L., Bekaert, K., Hoffman, S., Vandermeersch, G., Cooreman, K., & Robbens, J. (2014). Quality assessment of the blue mussel (Mytilus edulis): Comparison between commercial and wild types. *Marine Pollution Bulletin*, *85*(1), 146–155.

Dehaut, A., Cassone, A. L., Frère, L., Hermabessiere, L., Himber, C., Rinnert, E., Rivière, G., Lambert, C., Soudant, P., Huvet, A., & Duflos, G. (2016). Microplastics in seafood: Benchmark protocol for their extraction and characterization. *Environmental Pollution*, *215*, 223–233.

Desforges, J. P. W., Galbraith, M., Dangerfield, N., & Ross, P. S. (2014). Widespread distribution of microplastics in subsurface seawater in the NE Pacific Ocean. *Marine Pollution Bulletin*, *79*(1–2), 94–99.

Devi, S. S., Sreedevi, A. V., & Kumar, A. B. (2020). First report of microplastic ingestion by the alien fish Pirapitinga (Piaractus brachypomus) in the Ramsar site Vembanad Lake, south India. *Marine Pollution Bulletin*, *160*, 111637.

Devriese, L. I., Van der Meulen, M. D., Maes, T., Bekaert, K., Paul-Pont, I., Frère, L., Robbens, J., & Vethaak, A. D. (2015). Microplastic contamination in brown shrimp (Crangon crangon, Linnaeus 1758) from coastal waters of the Southern North Sea and Channel area. *Marine Pollution Bulletin*, *98*(1–2), 179–187.

Digka, N., Tsangaris, C., Kaberi, H., Adamopoulou, A., & Zeri, C. (2018a). Microplastic abundance and polymer types in a Mediterranean environment. In Proceedings of the International Conference on Microplastic Pollution in the Mediterranean Sea (pp. 17–24). Springer, Cham.

Digka, N., Tsangaris, C., Torre, M., Anastasopoulou, A., & Zeri, C. (2018b). Microplastics in mussels and fish from the Northern Ionian Sea. *Marine Pollution Bulletin*, *135*, 30–40.

Dowarah, K., Patchaiyappan, A., Thirunavukkarasu, C., Jayakumar, S., & Devipriya, S. P. (2020). Quantification of microplastics using Nile Red in two bivalve species Perna viridis and Meretrix meretrix from three estuaries in Pondicherry, India and microplastic uptake by local communities through bivalve diet. *Marine Pollution Bulletin*, *153*, 110982.

Dubaish, F., & Liebezeit, G. (2013). Suspended microplastics and black carbon particles in the Jade system, southern North Sea. *Water, Air, & Soil Pollution*, *224*(2), 1–8.

Elkhatib, D., & Oyanedel-Craver, V. (2020). A critical review of extraction and identification methods of microplastics in wastewater and drinking water. *Environmental Science & Technology*, *54*(12), 7037–7049.

Enders, K., Lenz, R., Beer, S., & Stedmon, C. A. (2017). Extraction of microplastic from biota: Recommended acidic digestion destroys common plastic polymers. *ICES Journal of Marine Science*, *74*(1), 326–331.

Eriksen, M., Lebreton, L. C., Carson, H. S., Thiel, M., Moore, C. J., Borerro, J. C., Galgani, F., Ryan, P. G., & Reisser, J. (2014). Plastic pollution in the world's oceans: More than 5 trillion plastic pieces weighing over 250,000 tons afloat at sea. *PloS One*, *9*(12), e111913.

Fang, C., Zheng, R., Chen, H., Hong, F., Lin, L., Lin, H., Guo, H., Bailey, C., Segner, H., Mu, J., & Bo, J. (2019). Comparison of microplastic contamination in fish and bivalves from two major cities in Fujian province, China and the implications for human health. *Aquaculture*, *512*, 734322.

Flotron, V., Delteil, C., Padellec, Y., & Camel, V. (2005). Removal of sorbed polycyclic aromatic hydrocarbons from soil, sludge and sediment samples using the Fenton's reagent process. *Chemosphere*, *59*(10), 1427–1437.

Friesen, L. W., Granberg, M. E., Hassellöv, M., Gabrielsen, G. W., & Magnusson, K. (2019). An efficient and gentle enzymatic digestion protocol for the extraction of microplastics from bivalve tissue. *Marine pollution bulletin*, *142*, 129–134.

Ganesan, M., Nallathambi, G., & Srinivasalu, S. (2019). Fate and transport of microplastics from water sources. *Current Science*, *117*(11), 00113891.

Gniadek, M., & Dąbrowska, A. (2019). The marine nano-and microplastics characterisation by SEM-EDX: the potential of the method in comparison with various physical and chemical approaches. *Marine Pollution Bulletin*, *148*, 210–216.

Goswami, P., Vinithkumar, N. V., & Dharani, G. (2020). First evidence of microplastics bioaccumulation by marine organisms in the Port Blair Bay, Andaman Islands. *Marine Pollution Bulletin*, *155*, 111163.

Goswami, P., Vinithkumar, N. V., & Dharani, G. (2021). Microplastics particles in seafloor sediments along the Arabian Sea and the Andaman Sea continental shelves: First insight on the occurrence, identification, and characterization. *Marine Pollution Bulletin*, *167*, 112311.

Gündoğdu, S. (2018). Contamination of table salts from Turkey with microplastics. *Food Additives & Contaminants: Part A*, *35*(5), 1006–1014.

Hanvey, J. S., Lewis, P. J., Lavers, J. L., Crosbie, N. D., Pozo, K., & Clarke, B. O. (2017). A review of analytical techniques for quantifying microplastics in sediments. *Analytical Methods*, *9*(9), 1369–1383.

Hossain, M. S., Sobhan, F., Uddin, M. N., Sharifuzzaman, S. M., Chowdhury, S. R., Sarker, S., & Chowdhury, M. S. N. (2019). Microplastics in fishes from the Northern Bay of Bengal. *Science of the Total Environment*, *690*, 821–830.

Hurley, R. R., Lusher, A. L., Olsen, M., & Nizzetto, L. (2018). Validation of a method for extracting microplastics from complex, organic-rich, environmental matrices. *Environmental Science & Technology*, *52*(13), 7409–7417.

Imhof, H. K., Schmid, J., Niessner, R., Ivleva, N. P., & Laforsch, C. (2012). A novel, highly efficient method for the separation and quantification of plastic particles in sediments of aquatic environments. *Limnology and Oceanography: Methods*, *10*(7), 524–537.

ICES (International Council for The Exploration of the Sea) (2015). OSPAR request on development of a common monitoring protocol for plastic particles in fish stomachs and selected shellfish on the basis of existing fish disease surveys. In *ICES Advice 2015, Book 1* (pp. 1–6).

IUCN (2018). *Overview of Marine Plastics: Report. Issues Brief.* IUCN.

James, K., Vasant, K., Padua, S., Gopinath, V., Abilash, K. S., Jeyabaskaran, R., Babu, A., & John, S. (2020). An assessment of microplastics in the ecosystem and selected commercially important fishes off Kochi, south eastern Arabian Sea, India. *Marine Pollution Bulletin, 154,* 111027.

Jung, H., Baek, M., D'Elia, K. P., Boisvert, C., Currie, P. D., Tay, B. H., ... & Dasen, J. S. (2018). The ancient origins of neural substrates for land walking. *Cell, 172*(4), 667–682.

Jungnickel, H., Pund, R., Tentschert, J., Reichardt, P., Laux, P., Harbach, H., & Luch, A. (2016). Time-of-flight secondary ion mass spectrometry (ToF-SIMS)-based analysis and imaging of polyethylene microplastics formation during sea surf simulation. *Science of the Total Environment, 563,* 261–266.

Käppler, A., Windrich, F., Löder, M. G., Malanin, M., Fischer, D., Labrenz, M., Eichhorn, K. J., & Voit, B. (2015). Identification of microplastics by FTIR and Raman microscopy: A novel silicon filter substrate opens the important spectral range below 1300 cm– 1 for FTIR transmission measurements. *Analytical and Bioanalytical Chemistry, 407*(22), 6791–6801.

Karami, A., Golieskardi, A., Choo, C. K., Larat, V., Galloway, T. S., & Salamatinia, B. (2017a). The presence of microplastics in commercial salts from different countries. *Scientific Reports, 7*(1), 1–11.

Karami, A., Golieskardi, A., Choo, C. K., Romano, N., Ho, Y. B., & Salamatinia, B. (2017b). A high-performance protocol for extraction of microplastics in fish. *Science of the Total Environment, 578,* 485–494.

Karami, A., Golieskardi, A., Choo, C. K., Larat, V., Karbalaei, S., & Salamatinia, B. (2018). Microplastic and mesoplastic contamination in canned sardines and sprats. *Science of the Total Environment, 612,* 1380–1386.

Karbalaei, S., Golieskardi, A., Hamzah, H. B., Abdulwahid, S., Hanachi, P., Walker, T. R., & Karami, A. (2019). Abundance and characteristics of microplastics in commercial marine fish from Malaysia. *Marine Pollution Bulletin, 148,* 5–15.

Karlsson, T. M., Vethaak, A. D., Almroth, B. C., Ariese, F., van Velzen, M., Hassellöv, M., & Leslie, H. A. (2017). Screening for microplastics in sediment, water, marine invertebrates and fish: Method development and microplastic accumulation. *Marine Pollution Bulletin, 122*(1–2), 403–408.

Karuppasamy, P. K., Ravi, A., Vasudevan, L., Elangovan, M. P., Mary, P. D., Vincent, S. G., & Palanisami, T. (2020). Baseline survey of micro and mesoplastics in the gastro-intestinal tract of commercial fish from Southeast coast of the Bay of Bengal. *Marine Pollution Bulletin, 153,* 110974.

Khoironi, A., & Anggoro, S. (2018, March). The existence of microplastic in Asian green mussels. *IOP Conference Series: Earth and Environmental Science, 131*(1), 012050. IOP Publishing.

Koelmans, A. A., Nor, N. H. M., Hermsen, E., Kooi, M., Mintenig, S. M., & De France, J. (2019). Microplastics in freshwaters and drinking water: Critical review and assessment of data quality. *Water Research, 155,* 410–422.

Kolandhasamy, P., Su, L., Li, J., Qu, X., Jabeen, K., & Shi, H. (2018). Adherence of microplastics to soft tissue of mussels: A novel way to uptake microplastics beyond ingestion. *Science of the Total Environment, 610,* 635–640.

Kosuth, M., Mason, S. A., & Wattenberg, E. V. (2018). Anthropogenic contamination of tap water, beer, and sea salt. *PloS One*, *13*(4), e0194970.

Kumar, V. E., Ravikumar, G., & Jeyasanta, K. I. (2018). Occurrence of microplastics in fishes from two landing sites in Tuticorin, South east coast of India. *Marine Pollution Bulletin*, *135*, 889–894.

Leslie, H. A., Brandsma, S. H., Van Velzen, M. J. M., & Vethaak, A. D. (2017). Microplastics en route: Field measurements in the Dutch river delta and Amsterdam canals, wastewater treatment plants, North Sea sediments and biota. *Environment International*, *101*, 133–142.

Li, B., Su, L., Zhang, H., Deng, H., Chen, Q., & Shi, H. (2020a). Microplastics in fishes and their living environments surrounding a plastic production area. *Science of the Total Environment*, 138662.

Li, J., Green, C., Reynolds, A., Shi, H., & Rotchell, J. M. (2018). Microplastics in mussels sampled from coastal waters and supermarkets in the United Kingdom. *Environmental pollution*, *241*, 35–44.

Li, J., Qu, X., Su, L., Zhang, W., Yang, D., Kolandhasamy, P., Li, D., & Shi, H. (2016). Microplastics in mussels along the coastal waters of China. *Environmental Pollution*, *214*, 177–184.

Li, J., Yang, D., Li, L., Jabeen, K., & Shi, H. (2015). Microplastics in commercial bivalves from China. *Environmental Pollution*, *207*, 190–195.

Liebezeit, G., & Liebezeit, E. (2013). Non-pollen particulates in honey and sugar. *Food Additives & Contaminants: Part A*, *30*(12), 2136–2140.

Liebezeit, G. & Liebezeit, E., (2015). Origin of synthetic particles in honeys. *Polish Journal of Food and Nutrition Sciences*, *65*(2), 143–147.

Löder, M. G., Imhof, H. K., Ladehoff, M., Löschel, L. A., Lorenz, C., Mintenig, S., Piehl, S., Primpke, S., Schrank, I., Laforsch, C., & Gerdts, G. (2017). Enzymatic purification of microplastics in environmental samples. *Environmental Science & Technology*, *51*(24), 14283–14292.

Lu, Y., Zhang, Y., Deng, Y., Jiang, W., Zhao, Y., Geng, J., Ding, L., & Ren, H. (2016). Uptake and accumulation of polystyrene microplastics in zebrafish (Danio rerio) and toxic effects in liver. *Environmental Science & Technology*, *50*(7), 4054–4060.

Lusher, A. L., Munno, K., Hermabessiere, L., & Carr, S. (2020). Isolation and extraction of microplastics from environmental samples: An evaluation of practical approaches and recommendations for further harmonization. *Applied Spectroscopy*, *74*(9), 1049–1065.

Lusher, A. L., O'Donnell, C., Officer, R., & O'Connor, I. (2016). Microplastic interactions with North Atlantic mesopelagic fish. *ICES Journal of Marine Science*, *73*(4), 1214–1225.

Lv, L., Qu, J., Yu, Z., Chen, D., Zhou, C., Hong, P., Sun, S., & Li, C. (2019). A simple method for detecting and quantifying microplastics utilizing fluorescent dyes-Safranine T, fluorescein isophosphate, Nile red based on thermal expansion and contraction property. *Environmental Pollution*, *255*, 113283.

Maes, T., Jessop, R., Wellner, N., Haupt, K., & Mayes, A. G. (2017). A rapid-screening approach to detect and quantify microplastics based on fluorescent tagging with Nile Red. *Scientific Reports*, *7*(1), 1–10.

Martinelli, J. C., Phan, S., Luscombe, C. K., & Padilla-Gamino, J. L. (2020). Low incidence of microplastic contaminants in Pacific oysters (Crassostrea gigas Thunberg) from the Salish Sea, USA. *Science of the Total Environment*, *715*, 136826.

Mason, S. A., Welch, V. G., & Neratko, J. (2018). Synthetic polymer contamination in bottled water. *Frontiers in Chemistry, 6,* 407.

Masura, J., Baker, J., Foster, G., & Arthur, C. (2015). *Laboratory Methods for the Analysis of Microplastics in the Marine Environment: Recommendations for Quantifying Synthetic Particles in Waters and Sediments.*

Mathalon, A., & Hill, P. (2014). Microplastic fibers in the intertidal ecosystem surrounding Halifax Harbor, Nova Scotia. *Marine Pollution Bulletin,* 81(1), 69–79.

Miller, M. E., Kroon, F. J., & Motti, C. A. (2017). Recovering microplastics from marine samples: A review of current practices. *Marine Pollution Bulletin, 123*(1–2), 6–18.

Morgana, S., Ghigliotti, L., Estévez-Calvar, N., Stifanese, R., Wieckzorek, A., Doyle, T., Christiansen, J. S., Faimali, M., & Garaventa, F. (2018). Microplastics in the Arctic: A case study with sub-surface water and fish samples off Northeast Greenland. *Environmental Pollution, 242,* 1078–1086.

Munno, K., Helm, P. A., Jackson, D. A., Rochman, C., & Sims, A. (2018). Impacts of temperature and selected chemical digestion methods on microplastic particles. *Environmental Toxicology and Chemistry, 37*(1), 91–98.

Murphy, C. L. (2018). *A Comparison of Microplastics in Farmed and Wild Shellfish Near Vancouver Island and Potential Implications for Contaminant Transfer to Humans* (Doctoral dissertation, Royal Roads University, Canada).

Naeeji, N., Rafeei, M., Azizi, H., Hashemi, M., & Eslami, A. (2020). Data on the microplastics contamination in water and sediments along the Haraz River estuary, Iran. *Data in Brief, 32,* 106155.

Nie, H., Wang, J., Xu, K., Huang, Y., & Yan, M. (2019). Microplastic pollution in water and fish samples around Nanxun Reef in Nansha Islands, South China Sea. *Science of the Total Environment, 696,* 134022.

Norén, F. (2008). Small plastic particles in coastal Swedish waters (p. 11). Lysekil: KIMO Sweden.

Nuelle, M. T., Dekiff, J. H., Remy, D., & Fries, E. (2014). A new analytical approach for monitoring microplastics in marine sediments. *Environmental Pollution, 184,* 161–169.

Oßmann, B. E., Sarau, G., Holtmannspötter, H., Pischetsrieder, M., Christiansen, S. H., & Dicke, W. (2018). Small-sized microplastics and pigmented particles in bottled mineral water. *Water Research, 141,* 307–316.

Oßmann, B., Schymanski, D., Ivleva, N. P., Fischer, D., Fischer, F., Dallmann, G., & Welle, F. (2019). Comment on "exposure to microplastics (< 10 μm) associated to plastic bottles mineral water consumption: The first quantitative study by Zuccarello et al. [Water Research 157 (2019) 365–371]". *Water Research, 162,* 516–517.

Pan, Z., Guo, H., Chen, H., Wang, S., Sun, X., Zou, Q., Zhang, Y., Lin, H., Cai, S., & Huang, J. (2019). Microplastics in the Northwestern Pacific: Abundance, distribution, and characteristics. *Science of the Total Environment, 650,* 1913–1922.

Pfohl, P., Roth, C., Meyer, L., Heinemeyer, U., Gruendling, T., Lang, C., Nestle, N., Hofmann, T., Wohlleben, W., & Jessl, S. (2021). Microplastic extraction protocols can impact the polymer structure. *Microplastics and Nanoplastics, 1*(1), 1–13.

Phuong, N. N., Poirier, L., Pham, Q. T., Lagarde, F., & Zalouk-Vergnoux, A. (2018a). Factors influencing the microplastic contamination of bivalves from the French Atlantic coast: Location, season and/or mode of life?. *Marine Pollution Bulletin, 129*(2), 664–674.

Phuong, N. N., Zalouk-Vergnoux, A., Kamari, A., Mouneyrac, C., Amiard, F., Poirier, L., & Lagarde, F. (2018b). Quantification and characterization of microplastics in blue mussels (Mytilus edulis): Protocol setup and preliminary data on the contamination of the French Atlantic coast. *Environmental Science and Pollution Research*, 25(7), 6135–6144.

Pivokonsky, M., Cermakova, L., Novotna, K., Peer, P., Cajthaml, T., & Janda, V. (2018). Occurrence of microplastics in raw and treated drinking water. *Science of the Total Environment*, 643, 1644–1651.

Plastics Europe (2019). *Plastics: The Facts*. Retrieved from. https://www.plasticseurope.org/en/resources/market-data.

Pozo, K., Gomez, V., Torres, M., Vera, L., Nuñez, D., Oyarzún, P., Mendoza, G., Clarke, B., Fossi, M. C., Baini, M., & Přibylová, P. (2019). Presence and characterization of microplastics in fish of commercial importance from the Biobío region in central Chile. *Marine Pollution Bulletin*, 140, 315–319.

Prata, J. C., Alves, J. R., da Costa, J. P., Duarte, A. C., & Rocha-Santos, T. (2020a). Major factors influencing the quantification of Nile Red stained microplastics and improved automatic quantification (MP-VAT 2.0). *Science of the Total Environment*, 719, 137498.

Prata, J. C., Castro, J. L., da Costa, J. P., Duarte, A. C., Rocha-Santos, T., & Cerqueira, M. (2020b). The importance of contamination control in airborne fibers and microplastic sampling: Experiences from indoor and outdoor air sampling in Aveiro, Portugal. *Marine Pollution Bulletin*, 159, 111522.

Prata, J. C., da Costa, J. P., Girão, A. V., Lopes, I., Duarte, A. C., & Rocha-Santos, T. (2019). Identifying a quick and efficient method of removing organic matter without damaging microplastic samples. *Science of the Total Environment*, 686, 131–139.

Prata, J. C., Paço, A., Reis, V., da Costa, J. P., Fernandes, A. J. S., da Costa, F. M., Duarte, A. C., & Rocha-Santos, T. (2020e). Identification of microplastics in white wines capped with polyethylene stoppers using micro-Raman spectroscopy. *Food Chemistry*, 331, 127323.

Primpke, S., Fischer, M., Lorenz, C., Gerdts, G., & Scholz-Böttcher, B. M. (2020). Comparison of pyrolysis gas chromatography/mass spectrometry and hyperspectral FTIR imaging spectroscopy for the analysis of microplastics. *Analytical and Bioanalytical Chemistry*, 412(30), 8283–8298.

Prudent, E., & Raoult, D. (2019). Fluorescence in situ hybridization, a complementary molecular tool for the clinical diagnosis of infectious diseases by intracellular and fastidious bacteria. *FEMS Microbiology Reviews*, 43(1), 88–107.

Qiu, Q., Tan, Z., Wang, J., Peng, J., Li, M., & Zhan, Z. (2016). Extraction, enumeration and identification methods for monitoring microplastics in the environment. *Estuarine, Coastal and Shelf Science*, 176, 102–109.

Qu, X., Su, L., Li, H., Liang, M., & Shi, H. (2018). Assessing the relationship between the abundance and properties of microplastics in water and in mussels. *Science of the Total Environment*, 621, 679–686.

Railo, S., Talvitie, J., Setälä, O., Koistinen, A., & Lehtiniemi, M. (2018). Application of an enzyme digestion method reveals microlitter in Mytilus trossulus at a wastewater discharge area. *Marine Pollution Bulletin*, 130, 206–214.

Ramasamy, E. V., Sruthi, S. N., Harit A. K., & Babu, N. (2019). Microplastics in human consumption: Table salt contaminated with microplastics. In S. Babel, A. Haarstrick, M. S. Babel, A. Sharp (Eds.), *Microplastics in the Water Environment* (pp. 74–80). Cuvillier Verlag. (ISBN 978-3-7369-7089-2 eISBN 978-3-7369-6089-3)

Renzi, M., & Blašković, A. (2018). Litter & microplastics features in table salts from marine origin: Italian versus Croatian brands. *Marine Pollution Bulletin, 135*, 62–68.

Renzi, M., Guerranti, C., & Blašković, A. (2018). Microplastic contents from maricultured and natural mussels. *Marine pollution bulletin, 131*, 248–251.

Roch, S., & Brinker, A. (2017). Rapid and efficient method for the detection of microplastic in the gastrointestinal tract of fishes. *Environmental Science & Technology, 51*(8), 4522–4530.

Rochman, C. M. (2015). The complex mixture, fate and toxicity of chemicals associated with plastic debris in the marine environment. In *Marine Anthropogenic Litter* (pp. 117–140). Springer.

Saliu, F., Montano, S., Garavaglia, M. G., Lasagni, M., Seveso, D., & Galli, P. (2018). Microplastic and charred microplastic in the Faafu Atoll, Maldives. *Marine Pollution Bulletin, 136*, 464–471.

Santana, M. F. M., Ascer, L. G., Custódio, M. R., Moreira, F. T., & Turra, A. (2016). Microplastic contamination in natural mussel beds from a Brazilian urbanized coastal region: Rapid evaluation through bioassessment. *Marine Pollution Bulletin, 106*(1–2), 183–189.

Sathish, M. N., Jeyasanta, I., & Patterson, J. (2020). Microplastics in salt of Tuticorin, southeast coast of India. *Archives of Environmental Contamination and Toxicology, 79*(1), 111–121.

Scheurer, M., & Bigalke, M. (2018). Microplastics in Swiss floodplain soils. *Environmental Science & Technology, 52*(6), 3591–3598.

Schymanski, D., Goldbeck, C., Humpf, H. U., & Fürst, P. (2018). Analysis of microplastics in water by micro-Raman spectroscopy: Release of plastic particles from different packaging into mineral water. *Water Research, 129*, 154–162.

Seth, C. K., & Shriwastav, A. (2018). Contamination of Indian sea salts with microplastics and a potential prevention strategy. *Environmental Science and Pollution Research, 25*(30), 30122–30131.

Shruti, V. C., Jonathan, M. P., Rodriguez-Espinosa, P. F., & Rodríguez-González, F. (2019). Microplastics in freshwater sediments of atoyac river basin, puebla city, Mexico. *Science of the Total Environment, 654*, 154–163.

Shruti, V. C., Pérez-Guevara, F., Elizalde-Martínez, I., & Kutralam-Muniasamy, G. (2020). First study of its kind on the microplastic contamination of soft drinks, cold tea and energy drinks-Future research and environmental considerations. *Science of the Total Environment, 726*, 138580.

Slootmaekers, B., Carteny, C. C., Belpaire, C., Saverwyns, S., Fremout, W., Blust, R., & Bervoets, L. (2019). Microplastic contamination in gudgeons (Gobio gobio) from Flemish rivers (Belgium). *Environmental Pollution, 244*, 675–684.

Sruthy, S., & Ramasamy, E. V. (2017). Microplastic pollution in Vembanad Lake, Kerala, India: The first report of microplastics in lake and estuarine sediments in India. *Environmental Pollution, 222*, 315–322.

Su, L., Nan, B., Hassell, K. L., Craig, N. J., & Pettigrove, V. (2019). Microplastics biomonitoring in Australian urban wetlands using a common noxious fish (Gambusia holbrooki). *Chemosphere, 228*, 65–74.

Suaria, G., & Aliani, S. (2014). Floating debris in the Mediterranean Sea. *Marine Pollution Bulletin, 86*(1–2), 494–504.

Tagg, A. S., Harrison, J. P., Ju-Nam, Y., Sapp, M., Bradley, E. L., Sinclair, C. J., & Ojeda, J. J. (2017). Fenton's reagent for the rapid and efficient isolation of microplastics from wastewater. *Chemical Communications, 53*(2), 372–375.

Tamminga, M., Hengstmann, E., & Fischer, E. K. (2018). Microplastic analysis in the South Funen Archipelago, Baltic Sea, implementing manta trawling and bulk sampling. *Marine Pollution Bulletin*, *128*, 601–608.

Tanaka, K., & Takada, H. (2016). Microplastic fragments and microbeads in digestive tracts of planktivorous fish from urban coastal waters. *Scientific Reports*, *6*(1), 1–8.

Thompson, R. C., Olsen, Y., Mitchell, R. P., Davis, A., Rowland, S. J., John, A. W., … & Russell, A. E. (2004). Lost at sea: where is all the plastic? *Science*, *304*(5672), 838–838.

Toussaint, B., Raffael, B., Angers-Loustau, A., Gilliland, D., Kestens, V., Petrillo, M., Rio-Echevarria, I. M., & Van den Eede, G. (2019). Review of micro-and nano-plastic contamination in the food chain. *Food Additives & Contaminants: Part A*, *36*(5), 639–673.

UNEP (2019). *Our Planet Is Drowning in Plastic Pollution*. UNEP.

Van Cauwenberghe, L., & Janssen, C. R. (2014). Microplastics in bivalves cultured for human consumption. *Environmental Pollution*, *193*, 65–70.

Van Cauwenberghe, L., Claessens, M., Vandegehuchte, M. B., & Janssen, C. R. (2015a). Microplastics are taken up by mussels (Mytilus edulis) and lugworms (Arenicola marina) living in natural habitats. *Environmental Pollution*, *199*, 10–17.

Van Cauwenberghe, L., Devriese, L., Galgani, F., Robbens, J., & Janssen, C. R. (2015b). Microplastics in sediments: A review of techniques, occurrence and effects. *Marine Environmental Research*, *111*, 5–17.

Van Cauwenberghe, L., Vanreusel, A., Mees, J., & Janssen, C. R. (2013). Microplastic pollution in deep-sea sediments. *Environmental Pollution*, *182*, 495–499.

Vandermeersch, G., Van Cauwenberghe, L., Janssen, C. R., Marques, A., Granby, K., Fait, G., Kotterman, M. J., Diogène, J., Bekaert, K., Robbens, J., & Devriese, L. (2015). A critical view on microplastic quantification in aquatic organisms. *Environmental Research*, *143*, 46–55.

Viršek, M. K., Palatinus, A., Koren, Š., Peterlin, M., Horvat, P., & Kržan, A. (2016). Protocol for microplastics sampling on the sea surface and sample analysis. *JoVE (Journal of Visualized Experiments)*, *118*, e55161.

Wang, W., Ndungu, A. W., Li, Z., & Wang, J. (2017). Microplastics pollution in inland freshwaters of China: A case study in urban surface waters of Wuhan, China. *Science of the Total Environment*, *575*, 1369–1374.

Woodall, L. C., Sanchez-Vidal, A., Canals, M., Paterson, G. L., Coppock, R., Sleight, V., … & Thompson, R. C. (2014). The deep sea is a major sink for microplastic debris. *Royal Society Open Science*, *1*(4), 140317.

Yang, D., Shi, H., Li, L., Li, J., Jabeen, K., & Kolandhasamy, P. (2015). Microplastic pollution in table salts from China. *Environmental Science & Technology*, *49*(22), 13622–13627.

Yang, X., Bento, C. P., Chen, H., Zhang, H., Xue, S., Lwanga, E. H., … & Geissen, V. (2018). Influence of microplastic addition on glyphosate decay and soil microbial activities in Chinese loess soil. *Environmental Pollution*, *242*, 338–347.

Yuan, W., Liu, X., Wang, W., Di, M., & Wang, J. (2019). Microplastic abundance, distribution and composition in water, sediments, and wild fish from Poyang Lake, China. *Ecotoxicology and Environmental Safety*, *170*, 180–187.

Zhang, J., Zhang, C., Deng, Y., Wang, R., Ma, E., Wang, J., Bai, J., Wu, J., & Zhou, Y. (2019). Microplastics in the surface water of small-scale estuaries in Shanghai. *Marine Pollution Bulletin*, *149*, 110569.

Zhao, S., Danley, M., Ward, J. E., Li, D., & Mincer, T. J. (2017). An approach for extraction, characterization and quantitation of microplastic in natural marine snow using Raman microscopy. *Analytical Methods, 9*(9), 1470–1478.

Zhu, L., Wang, H., Chen, B., Sun, X., Qu, K., & Xia, B. (2019). Microplastic ingestion in deep-sea fish from the South China Sea. *Science of the Total Environment, 677*, 493–501.

Zobkov, M., & Esiukova, E. (2017). Microplastics in Baltic bottom sediments: quantification procedures and first results. *Marine Pollution Bulletin, 114*(2), 724–732.

Zobkov, M. B., Esiukova, E. E., Zyubin, A. Y., & Samusev, I. G. (2019). Microplastic content variation in water column: The observations employing a novel sampling tool in stratified Baltic Sea. *Marine Pollution Bulletin, 138*, 193–205.

Impact of Microplastics on Human Health

5

5.1 INTRODUCTION

As one among the emerging fields of study, microplastics needs more research in understanding their impacts on biota and human health. This is evident from the fact that in recent years plastic production has increased severalfold, as it is being used widely in the health sector, a wide range of industries (textile, automobile, electrical and electronics industries), the construction sector, food and beverage production and packaging, etc., which are all closely linked with items being used in the day-to-day lives of human beings. It is also predicted that to the current level of plastics production (~370 million tons in 2019) an extra 33 billion tons of plastics will be added in 2050 (Galloway, 2015). This astounding increase in plastic production also warns us about the huge quantity of plastic waste ending up in the environment. Except for the meager quantity of plastics being recycled – only 9% of the production as per UNEP (2015) – the rest goes to landfills, dump yards, aquatic bodies and ultimately the ocean. According to Jambeck et al. (2015), 4.8–12.7 million tons of plastic waste entered the ocean in 2010; it has been estimated that around 12 billion tons of plastic litter will be in landfills and the environment by 2050 if the business-as-usual (BAU) mode of consumption and waste management practices continue (UNEP, 2018). This alarming scenario of plastic wastes entering the environment in general and the ocean in particular is quite relevant to this chapter as seafood items and sea salt are important means by which microplastics enter human beings through consumption. The presence of MPs with polymers such as polyethylene (PE) polystyrene (PS), polyvinyl chloride (PVC), polypropylene (PP), polyethylene terephthalate (PET), etc. has been reported in human food items such as fishes, bivalves, salt, drinking water, beer and honey (Sathish et al., 2020; James et al., 2020; Feng et al., 2019; Naidu, 2019; Ramasamy et al., 2019; Zhang et al., 2019; Kosuth et al., 2018;

DOI: 10.1201/9781003201755-5

Seth and Shriwastav, 2018); more details on polymer types in human food items are given in Tables 2.1, 2.2 and 2.3 of Chapter 2.

In general, natural and synthetic polymers such as cellulose, natural gas, crude oil and coal are used to produce plastics through two main processes: polymerization and polycondensation. In order to add strength and improve the quality of the plastics, a few additives (phthalates, bisphenol A, polybrominated diphenyl ethers and metals or metalloids) are added besides different colorants. It is to be noted that some of these additives are endocrine disruptors or carcinogenic, thus becoming a matter of concern to human health (Revel et al., 2018). Additionally, when the plastic debris or microplastics remain in the soil or waterbodies for longer periods, they do adsorb many substances from the surroundings like heavy metals, pesticide residues, persistent organic pollutants (POPs) besides a cocktail of microbes including pathogens develop as biofilms on the surface of the microplastics (Iñiguez et al., 2017; Talvitie et al., 2015; Yang et al., 2015). Thus, the consumption of MP-contaminated food items by humans can lead to health issues due to: (1) the physical nature of the MPs, mainly their size and shape; (2) the leaching of additives from MPs, i.e. chemicals like phthalates, bisphenols, colorants, etc.; (3) the substances adsorbed on the surface of MPs and (4) the microbial consortia of the biofilm found on the surface of the MPs.

Several studies have reported the toxicological impacts of MPs on different organisms (Rubio et al., 2020), while reports on human health are limited (Oberbeckmann et al., 2015; Rios et al., 2007; Mato et al., 2001). In this context, an overview of studies reporting the physiological, functional and toxic impacts of MPs on human health is presented in this chapter.

5.2 ENTRY OF MICROPLASTICS INTO THE HUMAN BODY

Entry of MPs into the human body can be due to direct or indirect exposure to MPs. Three potential means of direct exposure to MPs are found in the literature: ingestion, inhalation and dermal contact (Figure 5.1). Among these, the predominant mode of MP intake is through ingestion. Consumption of seafood items contaminated with MPs has been identified as a major means of MP intake by humans; consumption of sea salt also leads to MP intake. Tap water, bottled water and beverages also serve as sources of MPs in the human body. A recent study has reported the entry of MPs through the consumption of vegetables and fruits (Guo et al., 2020). Inhalation of indoor and outdoor dust also favors MP entry into the human body; the MPs found in dust mostly originate

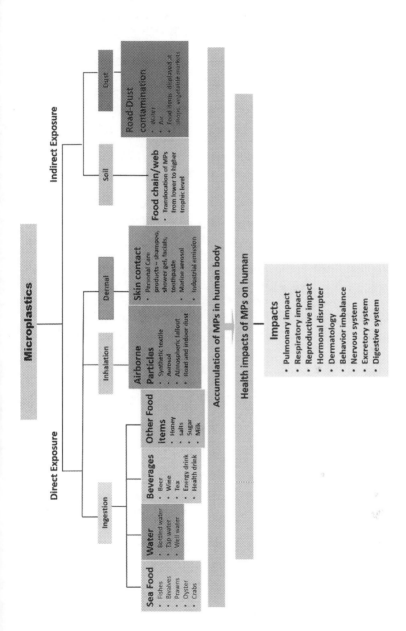

FIGURE 5.1 Pathways of microplastics entry into the human body and health impacts.

from synthetic textiles, aerosols and vehicle tires (Lehner et al, 2019; Stepleton, 2019). The dermal entry of MPs directly through the skin may be rare as the skin membrane is too fine to allow the MPs; however, MPs have been reported to enter through sweat glands, hair follicles and cuts or wounds in the skin.

Cox et al. (2019) have studied the quantity of MPs entering the human body through consumption (ingestion). Accordingly, the daily (average) concentration of MP intake by an adult male is 142 and for adult females it is 126; for male and female children the intake is 113 and 106 respectively. They have also calculated the concentration of MPs present in common human food items, for example, 1.48MPs/g of seafood, 0.11 MPs/g of salt, 0.44MPs/g of sugar, 94.37MPs/L of bottled water and 32.27MPs/L of alcohol. Through inhalation of air 9.80MPs/m^3 enter the human body, which means the daily average intake of MPs through inhalation is slightly higher than the entry through ingestion; for instance, an adult male and female inhale 170 and 132 MPs per day, respectively (Cox et al., 2019). Recent studies have reported MPs in human feces which in turn supports the view of humans' ingestion of MPs (Schwabl et al., 2019; Shruti et al., 2020).

Microplastics in the human body may cause harmful impacts; the impacts can be due to the MPs' entry leading to physiological and functional impacts on the human system/organs/tissues/cells; leaching out of additives and chemicals from the MPs may also cause harmful impacts on human health.

5.3 PHYSIOLOGICAL AND FUNCTIONAL IMPACTS ON HUMAN HEALTH

As mentioned earlier, studies are limited on the health impacts of MPs on human systems. The potential impacts reported in studies are oxidative stress, inflammatory responses, disruptions of gut microorganisms, functional impacts on the gastrointestinal tract, epidermis and lung failure (Wright and Kelly, 2017; Smith et al., 2018).

Tables 5.1 and 5.2 show the findings of the studies (*in vivo* and *in vitro*) conducted on the impact of MPs on the human body. The leaching of chemicals and additives from MPs affects the performance of human organs and can cause cell damage, necrosis, apoptosis, endocrine disruption, immune and reproductive disorders.

The size of the MPs is an important factor that determines the level of impact on human health. Once MPs reach the gut they will release or leach out additives, chemicals and adsorbed toxic substances, which in turn will cause physiological

TABLE 5.1   Studies on Potential Impacts of Microplastics on Human Health

IMPACTS OF MICROPLASTICS ON HUMAN ORGANS AND THEIR FUNCTION

AFFECTED ORGAN/ TISSUE/CELL	IMPACTS/DISEASES	REFERENCES
Intestinal epithelium cells	Hypersensitivity, immune response, hemolysis, cytotoxicity, apoptosis, oxidative stress	Thubagere and Reinhard, 2010; Stock et al., 2019; Schirinzi et al., 2017
Cells and cell division	Lower cell viability and cell cycle arrest (in the S phase of cell cycle), pro-apoptosis, change in cell morphology, cytotoxicity, decrease in metabolic activity, disturbances at the proliferation and cytoskeletal levels of cells	Xu et al., 2019; Goodman et al., 2021
Gastrointestinal tract and excretory system	Increased intestinal permeability, inflammatory bowel diseases	Cox et al., 2019; Schwabl et al., 2019; Schmidt et al., 2013
Lungs and skin	Lesions in the respiratory systems, inflammatory responses, reduction of ventilatory capacity, breathlessness, cough, inflammation, bio-persistence, respiratory systems, dyspnea, cancer, asthma, bronchitis	Gasperi et al., 2018; Catarino et al., 2018; Prata, 2018; Wright and Kelly, 2017; Dehghani et al., 2017
Placenta	Absorption of chemicals leached from plastic particles	Grafmueller et al., 2015
Immune cells and blood cells	Low cytotoxicity effect, stimulated immune system, enhanced potential hypersensitivity, increased levels of cytokines and histamines in PBMCs, raw 264.7 and HMC-1 cells	Hwang et al., 2019
Cerebral and epithelial cells	Cytotoxic effects, skin disease, nervous system issues	Schirinzi et al., 2017
Lungs and skin	Cytotoxic, inflammatory effects, reproductive toxicity, genotoxicity, mutagenicity, carcinogenicity, skin disease, respiratory issues	Gasperi et al., 2018; Dehghani et al., 2017

TABLE 5.2 Impacts due to Leaching of Additives, Chemicals and Heavy Metals from Microplastics on Human Health

Additives, chemicals and heavy metals	Affected organ/ tissue/cell	Impacts/diseases	References
Additives and chemicals (e.g. bisphenol A and phthalates)	Reproductive organs, neurons, lungs, intestines, hormones, skin	Genital malformations, infertility, learning disorders, autism spectrum disorders, endocrine disrupters, disrupting hormones, diabetes, obesity, cancers (breast, testes and prostate)	Campanale et al., 2020; Hahladakis, 2020
Heavy metals (e.g. Zn, Pb, Cr, Co, Cd, Ti, Br, Hg, As, Sn, etc.)	Reproductive organs, respiratory system, neurons, lungs, intestines, hormones, skin	Cellular and tissue damage, breast cancer, apoptosis, cytotoxicity, genotoxicity, mutagenic and carcinogenic effects, DNA methylation, gastrointestinal damage, vomiting, nausea, infertility, oxidative stress, cell damage, ulcer and neurological effects	Campanale et al., 2020; Godwill et al., 2019; Hahladakis, 2020; Gandamalla et al., 2019; Jan et al., 2015; Hansen et al., 2013

impacts. Further penetration of MPs into the organs or lungs is determined by the size of the MPs (Cox et al., 2019). MPs a few microns in size will enter the cells or lungs directly by cellular uptake; if they are larger in size (up to 130 microns) they can still enter tissues through paracellular uptake; those larger than 150 microns are not absorbed (Cox et al., 2019; Bouwmeester et al., 2015). The presence of MPs (20 particles per 10 g) in human feces with a size range of 50 to 500 μm indicates that higher-sized MPs are eliminated (Schwabl et al., 2019); thus it is evident that the human excretory system plays a role in removing a significant portion of MPs from the human body (Smith et al., 2018; Campanale et al., 2020).

5.3.1 Lung Damage

Microplastics of smaller dimensions, mostly synthetic fibers, entering the human respiratory system can cause inflammatory and cytotoxic effects along with respiratory distress (Hurley et al., 2016; Dehghani et al., 2017; Prata 2018; Rezaei et al., 2019; Dong et al., 2019). Genotoxic and cytotoxic effects on pulmonary epithelial cells have been noticed due to the presence of polystyrene particles of size 50 nm (Paget et al., 2015). Asthma-like bronchial reactions, interstitial fibrosis, inflammatory and fibrotic changes in the bronchial and peribronchial tissue (chronic bronchitis) are some of the effects of the entry of a synthetic particle, and such symptoms were noted in the staff of textile industries who work in close proximity to acrylic, polyester and nylon fibers (Campanale et al., 2020).

5.3.2 Skin Damage and Reproductive Effects

A study by the German Federal Institute for Risk Assessment (BfR, 2015) on health hazards due to the prolonged use of personal care products (face washes, toothpaste, hand cleansers, etc.) with MPs results in the absorption of PE and PP particles in the tissues leading to skin damage. MPs from toothpaste when swallowed unintentionally can enter into the gastrointestinal tract; such ingestions can lead to alterations in chromosomes causing obesity, infertility and cancer (GESAMP, 2015; Sharma and Chatterjee, 2017).

5.3.3 Toxic Chemicals and Additives Impact

The additives present in the MPs such as bisphenol A (BPA), phthalates, acrylamide, styrene, brominated flame retardants (BFR) and triclosan, when transferred to tissues, may lead to many health issues (Galloway, 2015). Many of these additives such as phthalates, BPA and polybrominated diphenyl ether (PBDEs) are endocrine-disrupting chemicals (McGrath et al., 2017; Serrano et al., 2014; Lu et al., 2013). Liver function changes, insulin resistance, damages to a developing fetus and neurological functions are some of the toxic effects of BPA (Srivastava and Godara, 2013). Phthalate esters may lead to abnormal sexual development and birth defects (Cheng et al., 2015). A few plastics like styrofoam contain carcinogenic chemicals like benzene and styrene; these chemicals are highly toxic if consumed and might damage the lungs, reproductive systems and nervous system (Yang et al., 2011).

Heavy metals adsorbed over the surface of MPs may also get leached out in the digestive tracts of human beings. Apart from the adsorbed heavy metals, a few plastics like vinyl window blinds and plastic jewelry have lead as plastic stabilizers added during production. Exposure to lead may increase the risk of hypertension and negatively affect the kidneys and nervous system in adults, while decreased IQ level and reduced growth are the effects in children (CIEL, 2019).

5.4 CONCLUSION

Microplastics' entry into the human system either through direct or indirect exposure results in many health impacts. The predominant means of entry of MPs into human body has been reported to be ingestion, i.e. through consumption of MP-contaminated food items (like fishes, prawns, crabs, salts, sugar, bottled water, honey, beer, etc.). However, inhalation of dust also leads to a substantial quantity of MPs' entry into the human body; other means are through direct contact on the skin by the use of personal care products like shampoos, cleansers, toothpaste, etc. The additives in the plastic particles, chemicals and heavy metals adsorbed over the MPs all together cause human health issues when the MPs gain entry into the human body. Some of the health issues reported in the literature due to MPs are inflammation, oxidative stress, necrosis, apoptosis, endocrine disruption, immune disorder and cancer. As the studies are limited in this field, there exists debate on the size of MPs that gain entry into human organs/cells, the subsequent fate and the impacts of MPs on the human body. Most of the studies report that MPs of less than 20 µm should be able to penetrate organs, and particles measuring 10 µm should be able to access all organs and cross the cell membranes. There also exist studies reporting MPs found in human stool indicating that larger-sized particles (50 to 500 µm) are removed from the human body by the excretory system. Available information on MPs' impact on human beings and other organisms is limited, hence more studies should be conducted in this area of research.

REFERENCES

BfR (2015). Germany's Federal Institute for Risk Assessment | Facts About BPA https://www.factsaboutbpa.org/study/germanys-federal-institute-for-risk-assessment/

Bouwmeester, H., Hollman, P. C. H., & Peters, R. J. B. (2015). Potential health impact of environmentally released micro- and nanoplastics in the human food production chain: experiences from nanotoxicology. *Environmental Science and Technology, 49*, 8932–8947.

Campanale, C., Massarelli, C., Savino, I., Locaputo, V., & Uricchio, V. F. (2020). A detailed review study on potential effects of microplastics and additives of concern on human health. *International Journal of Environmental Research and Public Health, 17*(4), 1212.

Chatterjee, S., & Sharma, S. (2019). Microplastics in our oceans and marine health. *Field Actions Science Reports. The Journal of Field Actions*, (Special Issue 19), 54–61.

Cheng, X., Ma, L., Xu, D., Cheng, H., Yang, G., & Luo, M. (2015). Mapping of phthalate esters in suburban surface and deep soils around a metropolis-Beijing, China. *Journal of Geochemical Exploration*, *155*, 56–61.

Catarino, A. I., Macchia, V., Sanderson, W. G., Thompson, R. C., & Henry, T. B. (2018). Low levels of microplastics (MP) in wild mussels indicate that MP ingestion by humans is minimal compared to exposure via household fibres fallout during a meal. *Environmental Pollution*, *237*, 675–684.

CIEL (2019). *Health: The Hidden Costs of a Plastic Planet*. Center for International Environmental Law Technical Report.

Cox, K. D., Covernton, G. A., Davies, H. L., Dower, J. F., Juanes, F., & Dudas, S. E. (2019). Human consumption of microplastics. *Environmental Science & Technology*, *53*(12), 7068–7074.

Dehghani, S., Moore, F. and Akhbarizadeh, R. (2017). Microplastic pollution in deposited urban dust, Tehran metropolis, Iran. *Environmental Science and Pollution Research*, *24*(25), 20360–20371.

Dong, T., Zhang, Y., Ma, P., Zhang, Y., Bernardini, P., Ding, M., ... & Zhang, Y. (2019). Charge measurement of cosmic ray nuclei with the plastic scintillator detector of DAMPE. *Astroparticle Physics*, *105*, 31–36.

Feng, Z., Zhang, T., Li, Y., He, X., Wang, R., Xu, J., & Gao, G. (2019). The accumulation of microplastics in fish from an important fish farm and mariculture area, Haizhou Bay, China. *Science of the Total Environment*, *696*, 133948.

Galloway, T. S. (2015). Micro-and nano-plastics and human health. In *Marine Anthropogenic Litter* (pp. 343–366). Springer.

GESAMP (2015). Sources, fate and effects of microplastics in the marine environment: A global assessment. In Kershaw, P. J. (Ed.), *(IMO/FAO/UNESCO-IOC/UNIDO/WMO/IAEA/UN/UNEP/UNDP Joint Group of Experts on the Scientific Aspects of Marine Environmental Protection). Rep Stud GESAMP No. 90*, p. 96.

Gandamalla, D., Lingabathula, H., & Yellu, N. (2019). Nano titanium exposure induces dose-and size-dependent cytotoxicity on human epithelial lung and colon cells. *Drug and Chemical Toxicology*, *42*(1), 24–34.

Gasperi, J., Wright, S. L., Dris, R., Collard, F., Mandin, C., Guerrouache, M., Langlois, V., Kelly, F. J., & Tassin, B. (2018). Microplastics in air: Are we breathing it in?. *Current Opinion in Environmental Science & Health*, *1*, 1–5.

Godwill, E. A., Ferdinand, P. U.; Nwalo, N. F., & Unachukwu, M. (2019). *Mechanism and Health Effects of Heavy Metal Toxicity in Humans. Poisoning in the Modern World-New Tricks for an Old Dog* (pp. 1–23). Intechopen.

Goodman, K. E., Hare, J. T., Khamis, Z. I., Hua, T., & Sang, Q. X. A. (2021). Exposure of human lung cells to polystyrene microplastics significantly retards cell proliferation and triggers morphological changes. *Chemical Research in Toxicology*, *34*(4), 1069–1081.

Grafmueller, S., Manser, P., Diener, L., Diener, P. A., Maeder-Althaus, X., Maurizi, L., Jochum, W., Krug, H. F., Buerki-Thurnherr, T., Von Mandach, U., & Wick, P. (2015). Bidirectional transfer study of polystyrene nanoparticles across the placental barrier in an ex vivo human placental perfusion model. *Environmental Health Perspectives*, *123*(12), 1280–1286.

Guo, J. J., Huang, X. P., Xiang, L., Wang, Y. Z., Li, Y. W., Li, H., Cai, Q. Y., Mo, C. H., & Wong, M. H. (2020). Source, migration and toxicology of microplastics in soil. *Environment International*, *137*, 105263.

Hurley, K., Reeves, E. P., Carroll, T. P., & McElvaney, N. G. (2016). Tumor necrosis factor-α driven inflammation in alpha-1 antitrypsin deficiency: A new model of pathogenesis and treatment. *Expert review of respiratory medicine,10*(2), 207–222.

Hahladakis, J. N. (2020). Delineating and preventing plastic waste leakage in the marine and terrestrial environment. *Environmental Science and Pollution Research*, *27*(11), 12830–12837.

Hansen, E., Nilsson, N. H., Lithner, D., & Lassen, C. (2013). *Hazardous Substances in Plastic Materials*. Klima- og forurensningsdirektoratet.

Hwang, J., Choi, D., Han, S., Choi, J., & Hong, J. (2019). An assessment of the toxicity of polypropylene microplastics in human derived cells. *Science of the Total Environment*, *684*, 657–669.

Iñiguez, M. E., Conesa, J. A., & Fullana, A. (2017). Microplastics in Spanish table salt. *Scientific Reports*, *7*(1), 1–7.

Jambeck, J. R., Geyer, R., Wilcox, C., Siegler, T. R., Perryman, M., Andrady, A., Narayan, R., & Law, K. L. (2015). Plastic waste inputs from land into the ocean. *Science*, *347*(6223), 768–771.

James, K., Vasant, K., Padua, S., Gopinath, V., Abilash, K. S., Jeyabaskaran, R., Babu, A., & John, S. (2020). An assessment of microplastics in the ecosystem and selected commercially important fishes off Kochi, south eastern Arabian Sea, India. *Marine Pollution Bulletin*, *154*, 111027.

Jan, A. T., Azam, M., Siddiqui, K., Ali, A., Choi, I., & Haq, Q. M. (2015). Heavy metals and human health: Mechanistic insight into toxicity and counter defense system of antioxidants. *International Journal of Molecular Sciences*, *16*(12), 29592–29630.

Kosuth, M., Mason, S. A., & Wattenberg, E. V. (2018). Anthropogenic contamination of tap water, beer, and sea salt. *PloS One*, *13*(4), p.e0194970.

Lehner, R., Weder, C., Petri-Fink, A., & Rothen-Rutishauser, B. (2019). Emergence of nanoplastic in the environment and possible impact on human health. *Environmental Science & Technology*, *53*(4), 1748–1765.

Liebezeit, G., & Liebezeit, E. (2015). Origin of synthetic particles in honeys. *Polish Journal of Food and Nutrition Sciences*, *65*(2), 143–147. https://doi.10.1515/pjfns-2015-0025.

Lu, L., Wan, Z., Luo, T., Fu, Z., & Jin, Y. (2018). Polystyrene microplastics induce gut microbiota dysbiosis and hepatic lipid metabolism disorder in mice. *Science of the Total Environment*, *631*, 449–458.

Mato, Y., Isobe, T., Takada, H., Kanehiro, H., Ohtake, C., & Kaminuma, T. (2001). Plastic resin pellets as a transport medium for toxic chemicals in the marine environment. *Environmental Science & Technology*, *35*(2), 318–324.

McGrath, T. J., Morrison, P. D., Ball, A. S., & Clarke, B. O. (2017). Detection of novel brominated flame retardants (NBFRs) in the urban soils of Melbourne, Australia. *Emerging Contaminants*, *3*(1), 23–31.

Naidu, S. A. (2019). Preliminary study and first evidence of presence of microplastics and colorants in green mussel, Perna viridis (Linnaeus, 1758), from southeast coast of India. *Marine Pollution Bulletin*, *140*, 416–422.

Oberbeckmann, S., Löder, M. G., & Labrenz, M. (2015). Marine microplastic-associated biofilms: A review. *Environmental Chemistry*, *12*(5), 551–562.

Paget, V., Dekali, S., Kortulewski, T., Grall, R., Gamez, C., Blazy, K., ..., & Lacroix, G. (2015). Specific uptake and genotoxicity induced by polystyrene nanobeads with distinct surface chemistry on human lung epithelial cells and macrophages. *PloS one, 10*(4), e0123297.

Plastic Euro (2019). *Plastics: The Facts 2019*. Plastics Europe.

Prata, J. C. (2018). Microplastics in wastewater: State of the knowledge on sources, fate and solutions. *Marine Pollution Bulletin, 129*(1), 262–265.

Ramasamy, E. V., Sruthi, S. N., Harit A. K., & Babu, N. (2019). Microplastics in human consumption: Table salt contaminated with microplastics. In S.Babel, A.Haarstrick, M. S.Babel, & A.Sharp (Eds.), *Microplastics in the Water Environment* (pp. 74–80). Cuvillier Verlag. (ISBN 978-3-7369-7089-2 eISBN 978-3-7369-6089-3).

Revel, M., Châtel, A., & Mouneyrac, C. (2018). Micro (nano) plastics: A threat to human health?. *Current Opinion in Environmental Science & Health, 1*, 17–23.

Rezaei, M., Riksen, M. J., Sirjani, E., Sameni, A., & Geissen, V. (2019). Wind erosion as a driver for transport of light density microplastics. *Science of the Total Environment, 669*, 273–281.

Rios, L. M., Moore, C., & Jones, P. R. (2007). Persistent organic pollutants carried by synthetic polymers in the ocean environment. *Marine Pollution Bulletin, 54*(8), 1230–1237.

Rubio, L., Marcos, R., & Hernández, A. (2020). Potential adverse health effects of ingested micro-and nanoplastics on humans. Lessons learned from in vivo and in vitro mammalian models. *Journal of Toxicology and Environmental Health, Part B, 23*(2), 51–68.

Sathish, M. N., Jeyasanta, I., & Patterson, J. (2020). Occurrence of microplastics in epipelagic and mesopelagic fishes from Tuticorin, Southeast coast of India. *Science of the Total Environment, 720*, 137614.

Schirinzi, G. F., Pérez-Pomeda, I., Sanchís, J., Rossini, C., Farré, M., & Barceló, D. (2017). Cytotoxic effects of commonly used nanomaterials and microplastics on cerebral and epithelial human cells. *Environmental Research, 159*, 579–587.

Schmidt, C., Lautenschlaeger, C., Collnot, E. M., Schumann, M., Bojarski, C., Schulzke, J. D., Lehr, C. M., & Stallmach, A. (2013). Nano-and microscaled particles for drug targeting to inflamed intestinal mucosa: A first in vivo study in human patients. *Journal of Controlled Release, 165*(2), 139–145.

Schwabl, P., Köppel, S., Königshofer, P., Bucsics, T., Trauner, M., Reiberger, T., & Liebmann, B. (2019). Detection of various microplastics in human stool: A prospective case series. *Annals of Internal Medicine, 171*, 453–457. https://doi.org/10.7326/M19-0618

Serrano, M., Barcenilla, F., & Limon, E. (2014). Nosocomial infections in long-term health care facilities. *Enfermedades Infecciosas y Microbiologia Clinica, 32*(3), 191–198.

Seth, C. K., & Shriwastav, A. (2018). Contamination of Indian sea salts with microplastics and a potential prevention strategy. *Environmental Science and Pollution Research, 25*(30), 30122–30131.

Smith, M., Love, D. C., Rochman, C. M., & Neff, R. A. (2018). Microplastics in seafood and the implications for human health. *Current Environmental Health Reports, 5*(3), 375–386.

Srivastava, R. K., & Godara, S. (2013). Use of polycarbonate plastic products and human health. Artigo em Inglês | IMSEAR | ID: sea-153816 https://pesquisa.bvsalud.org/portal/resource/pt/sea-153816

Shruti, V. C., Pérez-Guevara, F., Elizalde-Martínez, I., & Kutralam-Muniasamy, G. (2020). First study of its kind on the microplastic contamination of soft drinks, cold tea and energy drinks-Future research and environmental considerations. *Science of the Total Environment*, *726*, 138580.

Stapleton, P. A. (2019). Toxicological considerations of nano-sized plastics. *AIMS Environmental Science*, *6*(5), 367.

Stenmarck, Å., Belleza, E. L., Fråne, A., & Busch, N. (2017). *Hazardous substances in plastics:–ways to increase recycling.* Nordic Council of Minister, Feb 08.

Stock, V., Böhmert, L., Lisicki, E., Block, R., Cara-Carmona, J., Pack, L. K., Selb, R., Lichtenstein, D., Voss, L., Henderson, C. J., & Zabinsky, E. (2019). Uptake and effects of orally ingested polystyrene microplastic particles in vitro and in vivo. *Archives of Toxicology*, *93*(7), 1817–1833.

Talvitie, J., Heinonen, M., Pääkkönen, J. P., Vahtera, E., Mikola, A., Setälä, O., & Vahala, R. (2015). Do wastewater treatment plants act as a potential point source of microplastics? Preliminary study in the coastal Gulf of Finland, Baltic Sea. *Water Science and Technology*, *72*(9), 1495–1504.

Thubagere, A., & Reinhard, B. M. (2010). Nanoparticle-induced apoptosis propagates through hydrogen-peroxide-mediated bystander killing: Insights from a human intestinal epithelium in vitro model. *ACS Nano*, *4*(7), 3611–3622.

UNEP (2018). Single-use plastics: A roadmap for sustainability. 104 p. https://wedocs.unep.org/bitstream/handle/20.500.11822/25496/singleUsePlastic_sustainability.pdf

Vethaak, A. D., & Leslie, H. A. (2016). Plastic debris is a human health issue. *Environmental Science & Technology*, *50*(13), 6825–6826.

Wright, S. L., & Kelly, F. J. (2017). Plastic and human health: A micro issue?. *Environmental Science & Technology*, *51*(12), 6634–6647.

Xu, X., Wang, S., Gao, F., Li, J., Zheng, L., Sun, C., He, C., Wang, Z., & Qu, L. (2019). Marine microplastic-associated bacterial community succession in response to geography, exposure time, and plastic type in China's coastal seawaters. *Marine Pollution Bulletin*, *145*, 278–286.

Yang, C. Z., Yaniger, S. I., Jordan, V. C., Klein, D. J., & Bittner, G. D. (2011). Most plastic products release estrogenic chemicals: A potential health problem that can be solved. *Environmental Health Perspectives*, *119*(7), 989–996.

Yang, D., Shi, H., Li, L., Li, J., Jabeen, K., & Kolandhasamy, P. (2015). Microplastic pollution in table salts from China. *Environmental Science & Technology*, *49*(22), 13622–13627.

Zhang, Y., Gao, T., Kang, S., & Sillanpää, M. (2019). Importance of atmospheric transport for microplastics deposited in remote areas. *Environmental Pollution*, *254*, 112953.

Emerging Trends in Microplastic Pollution Management

6

Legislation, Policies and Mitigation Strategy

6.1 INTRODUCTION

As elaborated in the preceding chapters, a steep increase in plastic production has been noticed in the past few decades as a result of the huge demand. Plastics use in day-to-day human life is inevitable; there could be no second opinion on this. However, the enormous quantity of plastic waste generation and the poor plastic waste management practices existing in most countries lead to environmental issues threatening not only human life but the entire biota and ecosystems around the globe. The pathway of plastic litter reaching oceans starts from the terrestrial environment; transported to oceans by the rivers, streams and other waterways, plastics accumulate in oceans and form a significant proportion of marine debris. Around 80% of the plastics found in oceans originate from land. Thus plastics are highly pervasive and ubiquitous in distribution; polar waters, deep ocean sediments and lakes on high-range mountains are no exemptions. Consequent to the accumulation and persistence of plastic litter in the oceans, a number of harmful impacts on aquatic ecosystems, aquatic biota and human life have been documented. Microplastics are one among the off-shoots of issues that emerged recently due to plastic littering. As mentioned in Chapter 1, the plastics during their transport or after reaching the ocean break up into very small particles due to various factors. These small plastic fragments (< 5 mm in dimension) are known as microplastics (MPs), more specifically secondary microplastics. The other group of microplastics found in marine and terrestrial environments is intentionally made microbeads (< 5 mm) which are used in various personal care products and in industrial applications; these are known as primary microplastics. According to UNEP (2018), annually 4–12 million tons of plastic wastes enter the oceans; now that we can understand the magnitude of the issue, we can also realize that the mitigation or control of MP pollution should start from the source, and virtually nothing can be done in the ocean once the MPs reach it. Therefore, the policies or mitigation measures should address issues related to the source of MPs or plastics and strategies suitable for efficient plastic waste management.

6.2 LEGISLATION, POLICIES AND MITIGATION STRATEGY: A GLOBAL SCENARIO

To address the prevention and mitigation of MPs' impact on the environment and human health, several steps have been taken by different organizations globally such as the World Health Organization (WHO), the Group of Experts on the Scientific Aspects of Marine Environmental Protection (GESAMP), the United Nations Environmental Programme (UNEP), the National Oceanic and Atmospheric Administration (NOAA), the European Union (EU) and the Ministry of Environment Forest and Climate Change (MoEFCC), India. The summary of the statements of these organizations is given in Table 6.1.

6.2.1 Legislations and Policies: Global Scenario

Around the world, plastic pollution and related issues are becoming a serious matter of concern. Addressing this issue, several steps have been initiated globally and a wide range of legislations and policies focused on plastic production and management (littering and dumping), reducing, reusing, recycling and the use of biopolymers (biodegradable and compostable), etc., have been framed. To phase out and cut off the excessive usage of single-use plastics by 2022, the United Nations Environment Program (UNEP) has organized

TABLE 6.1 The Statements of Global Organizations on Prevention and Mitigation of Plastic Pollution

ORGANIZATION/AGENCIES	STATEMENT/POLICY
World Health Organization (WHO)	In 2019, Maria Neira, Director of the WHO Department of Public Health, Environment and Social Determinants of Health, explained the urgent need to study more about the health impact of microplastics on human beings as it is found everywhere including in our drinking water (*Microplastics in Drinking-Water*).
United Nations Environment Programme (UNEP) and National Oceanic and Atmospheric Administration (NOAA)	In 2018, the UNEP launched a global campaign to eradicate primary plastic sources of marine plastic debris like microplastics used in cosmetics and reduce the use of single-use plastics by 2022 (Clean Sea Campaign [UNEP Clean Sea, 2017]). They also launched a global framework to reduce marine plastic pollution and its impacts on human health, the environment and the economy under the Honolulu Strategy, 2011 (UNEP and NOAA, 2011).
European Union (EU)	Implemented various legislations related to microplastic contamination in several fields such as: • Water Framework Directive (WED) – EC, 2000 • Marine Strategy Framework Directive (MSFD) – EC, 2000 • Integrated Coastal Zone Management (ICZM) – EC, 2008 • European Strategy for Plastic – EC, 2018 • Common Fishing Policy (CFP) – EC, 2018
Ministry of Environment, Forest and Climate Change (MoEFCC), India	On the occasion of the 150th birthday of Mahatma Gandhi, India launched several programs to phase out single-use plastics by 2022: • Mother India Plastic-Free • Swachhta Hi Seva • Swachha Bharat Abhiyan • Awareness campaign

a global campaign (Llorca et al., 2020). In this context, global nations have adopted several legislations and policies as per their geographical location and issues; a brief description of this is given below

The United State of America (USA) launched a global framework to reduce marine plastic pollution and its impacts on human health, the environment and the economy under the Honolulu Strategy, 2011 (UNEP/NOAA, 2011). In 2015, then-President Barack Obama signed the Microbead-Free Waters Act of 2015 (H.R. 1321, 2015); accordingly, a complete phase-out of microplastic beads from personal care products has been scheduled for January 1, 2018. California was the first state in the USA to fully ban single-use plastic bags by passing the Assembly Bill (Prop 67 bill). San Francisco was the first city to impose a total ban on the sale of portable plastic water bottles (Plastic Pollution Coalition, 2016c). In 2017, the Save Our Seas Act (S.756, 2017) was introduced to reduce the marine debris in coastal areas and also globally, starting the Clean Sea Campaign in association with UNEP (UNEP Clean Sea, 2017).

The Canadian government introduced the Microbeads in Toiletries Regulations, 2017 (*Canada Gazette*, November 5, 2016) under the Canadian Environmental Protection Act, 1999 to completely ban microbeads in cosmetics products, as microbeads are on the list of hazardous pollutants formulated by the Canadian Environmental Protection Act, 1999. This also constituted regulations imposing a ban on manufacture as well as import and export of personal care products containing microbeads (*The Globe and Mail Newspaper*, 2016); exemption was given to non-prescription drugs or natural health products from 2018. However, from July 1, 2019, non-prescription drugs or natural health products with microbeads were also banned (Katyal et al., 2020).

In 2008, European Union (EU) members adopted the Marine Strategy Framework Directive (MSFD) in order to achieve "good environmental status" (GES) in the marine environment of European members by the year 2020. The EU has also directed the member states (27 member states) to monitor microplastic pollution and promote research under the Horizon 2020 program (Gago et al., 2018; Liu et al., 2018). The EU has also introduced several legislations related to microplastic pollution in relevant fields, such as the Common Fishing Policy (CFP) and the Water Framework Directive (WED) in the context of fishing regulation, in order to control chemicals and nutrients entering water bodies (EC, 2018a, 2000). It also recommended Integrated Coastal Zone Management (ICZM) and the Marine Strategy Framework Directive (MSFD) to protect all the coastal areas and marine waters of the EU (EC, 2000; EC, 2008). In addition, the EU (2018) adopted the European Strategy for Plastic (EC, 2018). The new strategy focused on the life cycle assessment of plastic (design used, production, recycling and economics) in the EU (EC, 2018). Recently, the European Commission formed a law named the Directive on Single-Use Plastics, 2019 to tackle ten single-use plastic items (cotton bud sticks, cutlery, plates, straws and stirrers, balloons and sticks for balloons, food containers, cups for beverages, beverage containers, cigarette butts, plastic bags, packets and wrappers, wet wipes and sanitary items), which are the most found items on Europe's beaches and to emphasize adopting/promoting alternatives of sustainable nature in place of these items. Under this law, the EU fixed the target to collect and recycle 90% of plastic bottles by 2029 (EC, 2019). A complete ban on the distribution of plastic bags of lightweight category in supermarkets has been imposed by France (Eastaugh, 2016).

The UK had a strategic ambition to handle all plastic waste (recyclable, reusable or compostable) by 2025 and also to work with a "target" to remove avoidable plastic waste by the end of 2042 and an "ambition" of zero avoidable waste by 2050. In 2018, the UK government formed the Resources and Waste Management Strategy, which contains various policies (consultation on reforming the UK packaging producer responsibility system, plastic packaging tax consultation, introducing a deposit return scheme [DRS] in England, etc.) targeting the reduction of plastic pollution. In addition, the UK government has also signed many international agreements to reduce plastic pollution in the marine environment. For example, the Commonwealth Clean Oceans Alliance also had a 2019 manifesto commitment to "ban" the export of plastic waste to non-Organization for Economic Co-operation and Development (non-OECD) countries (Smith, 2021).

England has published guidelines to ban single-use plastic items such as straws, cotton buds and stirrers from October 1, 2020 under the Environmental Protection (Plastic Straws, Cotton Buds and Stirrers) (England) Regulations 2020. The Netherlands, Austria, Belgium, Luxembourg and Sweden's governments also announced a ban on microbeads in cosmetics and personal care products by the end of the year 2017 (UNEP, 2015; UKDEFRA, 2016).

From April 2020, the city of Vancouver in British Columbia also imposed a ban on plastic straws and single-use items used in the food industry like styrofoam containers, disposable plates and cups under the strategic plan of its Overall Zero Waste 2040 vision (City of Vancouver, 2019).

Africa as the single continent with the highest number of countries stands on top to impose a total ban on the production and usage of plastic bags (Table 6.2). Half of the countries in Africa adopted nationwide ban on plastic bags between 2014 and 2017. Among African nations, Kenya is the one to impose the strictest plastic bags ban in the world. The Kenyan government announced heavy penalties for violation: fines of up to $38,000 and jail for up to four years. Some other African countries such as Jamaica, St. Kitts and Nevis, the Bahamas, St. Vincent and the Grenadines, Zambia and Belize have proposed to take steps on banning single-use plastic bags (UNEP, 2018).

TABLE 6.2 Summary of Steps Initiated to Eliminate/Reduce Single-Use Plastic, Microbeads and Styrofoam Products: A Global Scenario

CONTINENT	COUNTRY	ACTION TAKEN BY GOVERNMENT/ PRIVATE/NGO	TYPE OF ACTION	YEAR OF LAW IMPLEMENTED TO ELIMINATE/REDUCE SINGLE-USE PLASTIC
Europe	Denmark	Government – National	Levy	1994
	Ireland	Government – National	Levy	2002 and 2007 (with review)
	Luxembourg	Public–private agreement	Levy	2004
	Belgium	Government – National	Levy	2007
		Local – Wallonia	Ban	2016
		Local – Brussels Capital Region	Ban	2017
	Spain	Public–private agreement	Levy	2008
		Local – Andalusia	Levy	2011
		Local – Catalonia	Ban	2017
	Latvia	Government – National	Levy	2009
	Malta	Government – National	Levy	2009
	Romania	Government – National	Levy	2009
	Italy	Government – National	Ban	2011
			Levy	2018
	United Kingdom	Local – Wales	Levy	2011
		Government – National	Law	2017
	Bulgaria	Government – National	Levy	2011
	Hungary	Government – National	Levy	2012

(Continued)

TABLE 6.2 (CONTINUED) Summary of Steps Initiated to Eliminate/Reduce Single-Use Plastic, Microbeads and Styrofoam Products: A Global Scenario

CONTINENT	COUNTRY	ACTION TAKEN BY GOVERNMENT/ PRIVATE/NGO	TYPE OF ACTION	YEAR OF LAW IMPLEMENTED TO ELIMINATE/REDUCE SINGLE-USE PLASTIC
	Croatia	Government – National	Levy	2014
	Portugal	Government – National	Levy	2015
	Austria	Public–private agreement	Levy	2016
	Lithuania	Government – National	Levy	2016
	Finland	Public–private agreement	Levy	2016
	Germany	Public–private agreement	Levy or ban	2016
	Switzerland	Public–private agreement	Levy	2016
	Netherlands	Government – National	Levy	2016
	France	Government – National	Ban	2016
	Estonia	Government – National	Levy	2017
	Greece	Government – National	Levy	2018
	Cyprus	Government – National	Levy	2018
	Czech Republic	Government – National	Levy	2018
North and South America	Canada	Local – Leaf Rapids	Ban	2007
		Local – Wood Buffalo	Ban	2010
		Local – Thompson	Ban	2010
		Public–private agreement	Levy	2016
		Local – Montreal	Ban	2018

(Continued)

TABLE 6.2 (CONTINUED)   Summary of Steps Initiated to Eliminate/Reduce Single-Use Plastic, Microbeads and Styrofoam Products: A Global Scenario

CONTINENT	COUNTRY	ACTION TAKEN BY GOVERNMENT/ PRIVATE/NGO	TYPE OF ACTION	YEAR OF LAW IMPLEMENTED TO ELIMINATE/REDUCE SINGLE-USE PLASTIC
	Washington, DC	Local	Levy	2010
	Costa Rica	Government	Ban	2017
	New York City	Local	Ban	2015
		Government	Ban	2017
	California	Local	Ban and levy	2012
	San Francisco, CA	Local	Ban and levy	2012
	Antigua and Barbuda	Government – National	Ban	2016 and 2017
	Argentina	Local – Buenos Aires	Ban	2017
	Belize	Government – National	Ban	2018
	Brazil	Local – Rio de Janeiro	Ban	2009 and 2015
	Chile	Local – Punta	Ban	2014
			Bill	2017
	Colombia	Government – National	Ban	2017
	Chicago, IL	Local	Levy	2016
	Ecuador	Local – Galapagos	Ban	2015
	Guatemala	Local – San Pedro La Laguna and other cities	Ban	2017

(Continued)

TABLE 6.2 (CONTINUED) Summary of Steps Initiated to Eliminate/Reduce Single-Use Plastic, Microbeads and Styrofoam Products: A Global Scenario

CONTINENT	COUNTRY	ACTION TAKEN BY GOVERNMENT/ PRIVATE\NGO	TYPE OF ACTION	YEAR OF LAW IMPLEMENTED TO ELIMINATE/REDUCE SINGLE-USE PLASTIC
	Guyana	Government – National	Ban	2016
	Haiti	Government – National	Ban	2013
	Honduras	Local – Roatan, Utila, Gunanaja	Ban	2016
	Mexico	Local – Mexico City	Ban	2010
		Local – Queretaro		2018
	Panama	Government – National	Ban	2018
	St. Vincent and the Grenadines	Government – National	Ban	2017
	America Samoa	Local	Ban	2011
	Hawaii	Local	Ban	2011
	Austin, TX	Local	Ban	2013
	Texas	Local	Ban	2013
	England	Government – National	Ban	2020
	Northern Ireland	Local – Northern Ireland	Levy	2013
		Government – National	Ban	2020
	Wales	Government – National	Ban	2020
	Scotland	Local – Scotland	Levy	2014
UK		Government – National	Ban	2020

(Continued)

TABLE 6.2 (CONTINUED) Summary of Steps Initiated to Eliminate/Reduce Single-Use Plastic, Microbeads and Styrofoam Products: A Global Scenario

CONTINENT	COUNTRY	ACTION TAKEN BY GOVERNMENT/ PRIVATE/NGO	TYPE OF ACTION	YEAR OF LAW IMPLEMENTED TO ELIMINATE/REDUCE SINGLE-USE PLASTIC
Africa	South Africa	Government – National	Ban	2003
	Eritrea	Government – National	Ban	2005
	Tanzania	Government and local	Ban	2006
	Botswana	Government – National	Levy	2007
	Ethiopia	Government – National	Ban	2007
	Rwanda	Government – National	Ban	2008
	Uganda	Government – National	Ban	2009
	Morocco	Government – National	Ban	2009 and 2016
	Egypt	Local – Hurghada	Ban	2009
	Zimbabwe	Government – National	Ban	2010 and 2017
	Chad	Local – N'Djamena	Ban	2010
	Mali	Government – National	Ban	2012
	Mauritania	Government – National	Ban	2013
	Cote d'Ivoire	Government – National	Ban	2014
	Cameroon	Government – National	Ban	2014
	Niger	Government – National	Ban	2015
	Somalia	Local – Somaliland	Ban	2015
	Malawi	Government – National	Ban	2015

(Continued)

TABLE 6.2 (CONTINUED) Summary of Steps Initiated to Eliminate/Reduce Single-Use Plastic, Microbeads and Styrofoam Products: A Global Scenario

CONTINENT	COUNTRY	ACTION TAKEN BY GOVERNMENT/PRIVATE/NGO	TYPE OF ACTION	YEAR OF LAW IMPLEMENTED TO ELIMINATE/REDUCE SINGLE-USE PLASTIC
	The Gambia	Government – National	Ban	2015
	Burkina Faso	Government – National	Ban	2015
	Senegal	Government – National	Ban	2016
	Mozambique	Government – National	Ban	2016
	Mauritius	Government – National	Ban	2016
	Tunisia	Government – National	Ban	2017
	East Africa	Regional	Ban	2017
	Cape Verde	Government – National	Ban	2017
	Benin	Government – National	Ban	2018
Asia	Bangladesh	Government	Complete ban	2002
	Pakistan	Local – Punjab, Sindh, Khyber and Islamabad Capital Territory	Ban	2013, 2017 and 2018
	India	Government	Ban/partial ban	2016 and 2018
	China	Government – National	Ban and levy	2008 and 2009
			Levy/ban	2015
	Sri Lanka	Government – National	Ban	2017
	Malaysia	Local – Penang State	Levy	2011
	Mongolia	Government – National	Ban	2009

(Continued)

TABLE 6.2 (CONTINUED) Summary of Steps Initiated to Eliminate/Reduce Single-Use Plastic, Microbeads and Styrofoam Products: A Global Scenario

CONTINENT	COUNTRY	ACTION TAKEN BY GOVERNMENT/ PRIVATE/NGO	TYPE OF ACTION	YEAR OF LAW IMPLEMENTED TO ELIMINATE/REDUCE SINGLE-USE PLASTIC
	Thailand	Government – Public–private cooperation	Ban	2020
	Myanmar	Local – Mandalay	Ban	2009
	Nepal	Government – National	Ban	2011
	Indonesia	Local – 23 cities	Levy/ban	2016
	Israel	Government – National	Ban and levy	2017
	Japan	Government	Ban	2020
	Bhutan	Government – National	Ban	2009
	Maldives	Government – National	Ban	2021
	Vietnam	Government – National	Levy	2012
	Philippines	Local – Muntinlupa	Ban	2011
Oceania	Australia	Public–private agreement	Levy or ban	2003, 2009, 2011, 2013, 2016 and 2018
	New Zealand	Government	Partial ban	2018
	Papua New Guinea	Government – National	Ban	2016
	Fiji	Government – National	Levy	2017
	Vanuatu	Government – National	Ban	2018
	Marshall Islands	Government – National	Ban	2017
	Palau	Government – National	Ban	2017

Source: adapted from UNEP (2018).

In the Oceania/Australia continent, most of the states – Australia, Papua New Guinea, Fiji, Marshall Islands, Vanuatu and Palau – announced bans/levies on single-use plastic bags (Table 6.2), while the New Zealand government has announced a ban on certain plastic items in New Zealand from June 2018 (UNEP, 2018).

Among South Asian countries, Bangladesh was the first county – perhaps in the world – that imposed a ban on light-weight plastic bags in 2002, followed by Bhutan (2005), Afghanistan, Nepal and Sri Lanka (2011), Pakistan (2013), Maldives (2016), China and India (2018) (Chowdhury et al., 2021; UNEP, 2018).

In 2016, India formulated the Plastic Waste Management Rules to manage plastic waste. In continuation of this, in 2018 while hosting UN World Environment Day, it promised to phase out single-use plastics by 2022 under the theme of "Beat Plastic Pollution". To achieve its goal, India started a nationwide plastic campaign to raise awareness among the people about plastic pollution and its impact on the environment and human health. The Government of India also focused attention on India's Himalayan regions (IHR) across the 12 Himalayan states (Jammu and Kashmir, Uttarakhand, Himachal Pradesh, Sikkim, Nagaland, Arunachal Pradesh, Manipur, Tripura, Mizoram, Meghalaya, Assam Hills and West Bengal Hills) by organizing an annual program under the slogan of "The Himalayan Cleanup" (THC) by the Integrated Mountain Initiative (IMI) and Zero Waste Himalaya (ZWH). In 2011, the Indian government also introduced the Plastic Waste Management Legislation, 2011. Recently India announced the Single-Use Plastic (Regulation) Act, 2018. Its goal is to eliminate single-use plastic from the nation by the year 2022.

In 2005, Israel's government launched a "Clean Coast" program as a joint venture with local communities/authorities and school/youth movements to end up plastic pollution on Israeli beaches (Alkalay et al., 2007). In Taiwan, a foundation named the Society of Wilderness conducted a beach cleanup program on 26 beaches in Taiwan from 2010 to 2012 (Society of Wilderness, 2014; Kuo and Huang, 2014). The Nepali government imposed a nationwide ban on plastic bags less than 20 microns in thickness (MOEST, 2011; Bharadwaj, 2016). The government of Thailand also has announced a plan to collect a levy on plastic bags and ban single-use plastic in 2020. Similarly, Japan's government also initiated a ban on free plastic shopping bags in the year 2020, while the Philippines' government announced bills banning/limiting the usage of single-use plastics and microplastics. China (one of the world's biggest users of plastic) also declared to ban single-use plastics (SUPs) across the country by 2022. The president of the Maldives also declared the ban on SUPs from June 1, 2021, and fixed the goal to phase out SUPs by 2023 (PR, 2020).

In summary, more than 30 countries in Asia, Africa, Europe, the Americas and Oceania have partially or completely banned or imposed levies on the use of plastic bags (Dikgang et al., 2012; Gold et al., 2014; EU, 2014). Among these, Bangladesh was the first country to outlaw polythene bags in 2002 followed by other nations in Asia, Africa, Europe and the Americas (Bergmann et al., 2015, Ogunola et al., 2018).

6.2.2 Mitigation Strategy

To mitigate plastic pollution, several countries have proposed new legislation particularly to control plastic littering, for example, countries from the Pacific, Latin America, the Caribbean, Africa and Asia. Initiatives were also taken by several developed (Germany, Japan) and developing (South Africa) countries through adopting extended producer responsibility (EPR) for recycling of used PET bottles by manufacturers either voluntarily or by law. Indeed, several national and international social organizations/NGOs also play a role in this venture. For example, the Plastic Soup Foundation (2016) has pleaded with the industries and customers to completely eliminate the usage of microbeads in personal care products (PCP). In continuation of this, international personal care brands such as Livon and L'Oréal and Johnson and Johnson have committed to completely phasing off microbeads and microplastics from their PCP (Copeland, 2015). In 2017 IKEA, a homeware company produced their sustainability summary report (fall 2017) according to which 590,259 tons of waste were generated by their food chain, of which 83% was recycled or incinerated for energy recovery after the adaptation of EPR policies (INGKA, 2017).

Steps have also been taken by several companies to utilize and upscale plastic litter either generated by them or other companies through removing, recycling and reclaiming plastic litter from the marine environment. For example, Adidas in association with Parley for the Ocean in 2015, manufactured sneakers (running shoes) and clothing from plastic waste in the Maldives via using a zero-waste 3D printing technique and in 2017 sold one million shoes, which is equivalent to 14.3 tons of nylon net and 16.5 million plastic bottles (Rhodes, 2016; Kharpal, 2017; Greenstein, 2017). Some international clothing companies like Burco, Unifi, City Place and G-star Raw clothing have also taken steps to minimize or reduce the plastic pollution in aquatic water bodies under the Ocean Plastic Programs AIR: avoid, intercep, and redesign (Greenstein, 2017).

Another approach to mitigating plastic pollution in coastal areas is a beach cleanup program, globally organized by governmental bodies (universities/institutions) and non-governmental organizations (NGOs) under the flagship of International Coastal Cleanup (ICC) by the Ocean Conservancy, a United States–based NGO (UN, 2017). The aim of ICC is the collection of plastic debris (meso- and macroplastics) from the beaches and to raise awareness

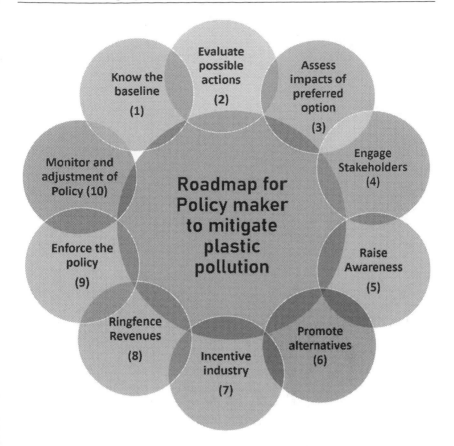

FIGURE 6.1 Roadmap suggested by UNEP (2018) for policymakers to mitigate single-use plastic pollution. (Source: adapted from UNEP, 2018.)

among the local vendors, local communities and tourists about plastic pollution and its harmful impacts.

In 2018, the UNEP suggested a ten-step roadmap (Figure 6.1) for policymakers to mitigate single-use plastic pollution. A summary of various steps taken by international organizations/countries for prevention and mitigation strategy is given in Table 6.2.

Some useful points for mitigating MPs pollution

1. Banning microplastic beads in cosmetics and personal care products like shampoos, scrubs, toothpaste, soaps and cleaners.

2. Upgrading water treatment plants for the removal of MP particles before release.
3. Modifying existing washing machines with efficient filter mechanisms for MP particles.
4. Innovating biodegradable plastics and using natural products in edible cutlery, food wrapping and straws.
5. Conducting more and more public awareness programs at the global, national and local levels, educating the public to reduce the use of plastic products and emphasizing proper management of plastic waste.

6.3 CONCLUSION

Global initiatives on various aspects of plastic waste and its management are quite inspiring. The strategies adopted by developed and less developed nations to mitigate and prevent plastic as well as microplastic pollution as mentioned at the beginning of this chapter focus on the source of the plastic waste or pollution. The emphasis given to recycling and reusing plastic waste through the extended producer responsibility (EPR) is the right approach to attain dual benefits of reducing plastic littering and a circular economy gain generating resources. More intense R&D on compostable plastics is also needed for the benefit of future generations. As the popular three "Rs" concept emphasizes, the successful management of plastic waste should always begin with the *reduction* of plastic use and reduction in plastic littering followed by the *reuse* of plastic items wherever possible and the *recycling* of plastic discards to enter the circular economy. Having said all this, it is needless to mention that no government can do it alone and win this challenge; only a partnership of the public, the producers (industries/corporations), NGOs and stringent government regulations can attain success in a sustainable manner.

REFERENCES

Alkalay, R., Pasternak, G., & Zask, A. (2007). Clean-coast index: A new approach for beach cleanliness assessment. *Ocean & Coastal Management, 50*(5–6), 352–362.

Bergmann, M., Gutow, L., & Klages, M. (2015). *Marine Anthropogenic Litter* (p. 447). Springer Nature.

Bharadwaj, B. (2016). Plastic bag ban in Nepal: Enforcement and effectiveness. Published by the South Asian Network for Development and Environmental Economics (SANDEE), ISSN 1893-1891; WP 111–16.

Chowdhury, H., Chowdhury, T., & Sait, S. M. (2021). Estimating marine plastic pollution from COVID-19 face masks in coastal regions. *Marine Pollution Bulletin*, *168*, 112419.

City of Vancouver (2019). Single-use item reduction strategy. https://vancouver.ca/green-vancouver/single-use-items.asp

Copeland, C. (2015). Microbeads: an emerging water quality issue. Library of Congress, Congressional Research Service. Retrieved from https://www.fas.org/sgp/crs/misc/IN10319

Dikgang, J., Leiman, A., & Visser, M. (2012). Analysis of the plastic-bag levy in South Africa. *Resources, Conservation and Recycling*, *66*, 59–65.

Eastaugh S., 2016. France becomes first country to ban plastic cups and plates. CNN; September 20, 2016. Available from: https://cnnphilippines.com/world/2016/09/20/france-ban-plastic-cupsplates.html.

EC (European Commission) (2000). Directive 2000/60/EC of the European Parliament and of the Council establishing a framework for the Community action in the field of water policy" or, for short, the EU Water Framework Directive (WFD) was finally adopted. https://ec.europa.eu/environment/water/water-framework/index_en.html

EC, 2008. On recycled plastic materials and articles intended to come into contact with foods and amending Regulation (EC) No 2023/2006 http://faolex.fao.org/docs/pdf/eur78350.pdf

EC (2018a). EU plastic strategy. https://ec.europa.eu/environment/strategy/plastics-strategy_en

EC (2018b). Common fisheries policy (CFP). https://ec.europa.eu/oceans-and-fisheries/policy/common-fisheries-policy-cfp_en#ecl-inpage-562

EC (2019). EU plastic strategy. https://ec.europa.eu/environment/strategy/plastics-strategy_en

EU (2014). Use of plastic bags: Agreement on phasing down. Council of the European Union. https://www.consilium.europa.eu/uedocs/cms_data/docs/pressdata/en/envir/145898.pdf

Gago, J., Frias, J., Filgueiras, A., Pedrotti, M. L., Suaria, G., & Tirelli, V. (2018). Standardised protocol for monitoring microplastics in seawater. Report number: D4. 1 BASEMAN Project. JPI-Oceans BASEMAN project.DOI: 10.13140/RG.2.2.14181.45282

Gold, M., Mika, K., Horowitz, C. and Herzog, M., 2014. Stemming the tide of plastic litter: A global action agenda. Tulane Environmental Law Journal, *27*, 165.

Greenstein, J. (2017). *Upcycled Ocean Plastic*. 2016; Available at: http://ocean.si.edu/ocean-news/upcycled-ocean-plastic.

H.R. 1321 (2015). *Microbead-Free Waters Act of 2015*. Government of the United States of America. https://www.cbo.gov/sites/default/files/114th-congress-2015-2016/costestimate/hr1321.pdf

INGKA (2017). *Sustainability Summary Report FY17*. Ikea. http://www.ikea.com/gb/en/doc/ikea-2017-ikea-group-sustainability-summary-report__1364488103883.pdf.

Katyal, D., Kong, E. and Villanueva, J., 2020. Microplastics in the environment: Impact on human health and future mitigation strategies. *Environmental Health Review*, *63*(1), 27–31.

Kharpal, A. (2017). Adidas sold 1 million shoes made out of ocean plastic in 2017. *CNBC*. 2018. https://www.cnbc.com/2018/03/14/adidas-sold-1-million-shoes-made-out-of-ocean-plastic-in-2017.html.

Kuo, F. J., & Huang, H. W. (20140. Strategy for mitigation of marine debris: Analysis of sources and composition of marine debris in northern Taiwan. *Marine Pollution Bulletin*, *83*(1), 70–78.

Liu, M., Lu, S., Song, Y., Lei, L., Hu, J., Lv, W., Zhou, W., Cao, C., Shi, H., Yang, X., & He, D. (2018). Microplastic and mesoplastic pollution in farmland soils in suburbs of Shanghai, China. *Environmental Pollution*, *242*, 855–862.

Llorca, M., Álvarez-Muñoz, D., Ábalos, M., Rodríguez-Mozaz, S., Santos, L. H., León, V. M., Campillo, J. A., Martínez-Gómez, C., Abad, E., & Farré, M. (2020). Microplastics in Mediterranean coastal area: Toxicity and impact for the environment and human health. Trends in Environmental Analytical Chemistry, *27*, e00090.

MOEST (2011). *Plastic Bag (Regulation and Reduction) Directive*. Ministry of Environment, Science and Techonology, Government of Nepal.

Ogunola, O. S., Onada, O. A. and Falaye, A. E., 2018. Mitigation measures to avert the impacts of plastics and microplastics in the marine environment (a review). *Environmental Science and Pollution Research*, *25*(10), 9293–9310.

Plastic Polution Coalition (2016c). The first American city to ban plastic bottles. http://www.plasticpollutioncoalition.org/pft/2016/2/19/thefirst-american-city-to-ban-plastic-water-bottles

PR (2020). President declares list of Single-use Plastics prohibited to import from June 1, 2021, Ref: 2020–497. https://presidency.gov.mv/Press/Article/24211

Rhodes, M. (2016). Adidas spins plastic from the ocean into awesome kicks. Adidas Spins Plastic from the Ocean into Awesome Kicks | WIRED.

S.756, 2017; Save our seas act of 2017 report of the committee on commerce, science and transportation pp 115-135 https://www.congress.gov/115/crpt/srpt135/CRPT-115srpt135.pdf

Smith, L. (2021). *Plastic Waste. House of Commons Library*, Briefing paper No. 08515, 12 May 2021. https://researchbriefings.files.parliament.uk/documents/CBP-8515/CBP-8515.pdf

Society of Wilderness (2014). *Reduce Plastic and Avoid the ('Age of Plastics')*. Society of Wilderness.

The Globe and Mail Newspaper (2016). Government of Canada labels microbeads 'toxic substance'. https://www.theglobeandmail.com/news/national/feds-label-microbeads-as-toxicsubstance/article30698903/

UKDEFRA 2016; Annual Report and Accounts 2015–16. https://assets.publishing.service.gov.uk/government/uploads/system/uploads/attachment_data/file/547295/defra-annual-report-2015-2016-print.pdf

UN (United Nations) (2017). *United Nations Convention on the Law of the Sea of 10 December 1982*. UN.

UNEP (2015). *Plastics in Cosmetics–Are We Polluting the Environment Through Our Personal Care?* Institute of Environmental Studies, VU University of Amsterdam.

UNEP (2018). *Single-use Plastics: A Roadmap for Sustainability.* United Nations Environment Programme. https://www.unep.org/resources/report/single-use -plastics-roadmap-sustainability

UNEP Clean Sea (2017). *Clean Seas campaign launches by United Nations Environment Programme (UNEP) in 2017.* https://www.cleanseas.org/about

UNEP/NOAA (2011). *The Honolulu Strategy: A Global Framework for Prevention and Management of Marine Debris.* The United Nations Environment Program (UNEP), Nairobi, Kenya, and National Oceanic and Atmospheric Administration (NOAA) Marine Debris Program. https://www.unenvironment.org/resources/ report/honolulustrategy

Index

Printed in the United States
by Baker & Taylor Publisher Services